T0187424

First published and distributed by
viction:workshop ltd.

viction:ary™

viction:workshop ltd.
Unit C, 7/F, Seabright Plaza,
9-23 Shell Street, North Point,
Hong Kong SAR
Website: www.victionary.com
Email: we@victionary.com
🔘 @victionworkshop
🔘 @victionworkshop
Bē @victionary
🔘 @victionary

Edited and produced by viction:workshop ltd.
Creative Direction: Victor Cheung
Design: Carol Chan, Cheryl Lai, Scarlet Ng
Editorial: Ynes Filleul, YL Lim
Coordination: Katherine Wong
Production: Bryan Leung

Jacket illustrations by Josh Evans, Esther Goh,
Lauren Humphrey and ELLAKOOKOO

ISBN 978-988-76844-4-2
Printed and bound in China

GET LOST

EXPLORE THE WORLD
THROUGH MAP ILLUSTRATIONS

PREFACE

When was the last time you found yourself in unfamiliar territory — whether you were at a crossroads in life or simply trying to situate yourself in a neighbourhood you have never been to?

It is human nature to fear the unknown. Our brain is programmed to constantly predict what happens next to prepare our body and mind for the potential dangers ahead in the best way possible. Studies have also shown that elements of unpredictability significantly increase our level of discomfort — even amid situations we have some form of control over. Although scientists continue to make great strides in explaining why uncertainty can be debilitating in hopes of helping those who suffer from anxiety disorders, the fact remains: our journey here on Earth involves treading on uncharted waters at some point or another, whether we like it or not.

We may have been told time and time again that failure to plan is planning to fail. While there is some truth to this saying when it comes to preparing for an important event for example, there is a sense of peace that comes from accepting that we will never truly know what lies in the future — and that any attempt to control the outcome of our decisions will often be futile.

Contrary to what we may think in the face of a challenge, being uncomfortable can do us more good than harm. Besides facilitating our growth, we also get to gain the courage to take more creative risks that aid us in solving problems and finding meaning in them. Plus, surprises can be fun! Rather than scare us into inaction, they can become a source of joy and excitement that allows us to make the most of our journey rather than focus purely on the destination itself.

The trick to feeling more comfortable outside our comfort zone? Practice, practice, practice. The more exposed we are to the unknown, the more open we can become to embrace all that it holds. Travelling can be a good way to get started, particularly to places we have yet to explore. Be they in the real world or within a fictitious one, new environments offer fresh inspiration and more room for our imagination to roam. On top of relieving stress, travelling gives us the opportunity to expand our mind as we try to understand and adapt to different settings, even if we are not too far away from home. There is also something freeing about not knowing who you will meet or where you will end up when you take a break from the expected and familiar.

So, throw your caution to the wind, jump in at the deep end — and GET LOST! In this book, you will find a myriad of maps that serve as gateways to new discoveries in person or from the comfort of your favourite armchair. Whether you already know of the locations featured or are seeing them for the first time, immerse yourself in the colourful cartographic wonders within these pages and let your mind wander as you diverge onto untrodden paths full of delightful possibilities — no navigation apps necessary!

TUSSEN DE BOGEN

WILLEM DE ZWIJGERLAAN

VAN HALLSTRAAT

Frederik Hendrik Plantsoen

WESTERSTRAAT

PRINSENGRACHT

SINGEL

National Monument

Royal Palace of Amsterdam

NASSAUKADE

MARNIXSTRAAT

HENDRIKSTRAAT

ADMIRAAL DE RUIJTERWEG

JAN VAN GALENSTRAAT

FREDERIK

Anne Frank House

RAADHUISSTRAAT

Dam Square

DE CLERCQSTRAAT

SINGEL

ROKIN

Jordaan District

Madame Tu

BILDERDIJKKADE

MARNIXSTRAAT

NASSAUKADE

Amsterdam Museum

AMSTE

LEIDSEGRACHT

Leidseplein

KINKERSTRAAT

VIJZELSTRAAT

PRINSENGRACHT

PC Hoofstraat

Rijksmuseum

OVERTOOM

VAN BAERLESTRAAT

i amsterdam

STADHOUDERSKADE

Van Gogh Museum

Vondelpark

Museumplein

Park Museumplein

HOBBEMAKADE

Sarphatit

DE LAIRESSESTRAAT

Olympic Stadium Amsterdam

CEINTUURBAAN

006

IJ

The Port of Amsterdam

Central Station

PIET HEINKADE

St. Nicolaaskerk

PRINS HENDRIKKADE

Chinatown

VALKENBURGERSTRAAT

Artis zoo

WEESPERSTRAAT

PLANTAGE MIDDENLAAN

Artis

Amstel River

SARPHATISTRAAT

Royal Theatre Carre

MAURITSKADE

SARPHATISTRAAT

WIBAUTSTRAAT

Oosterpark

Amstel

Wibauthius building

Amsterdam University of Applied Sciences

AMSTELDIJK

VAN WOUSTRAAT

Ajax Stadium

Fraijlemaborg building

VESA SAMMALISTO

Vesa Sammalisto, an award-winning illustrator residing and working in Helsinki, has an impressive client list that includes Sony, Twitter, Cartoon Network, BMW, Google, Qatar Airways, Monocle, and Wired. When not creating, he enjoys mountain biking through forests or riding the tracks at nearby bike parks.

Map of Amsterdam for Northumbria University Student Guide (↖)
Art Direction: Gardiner Richardson

Map of Newcastle for Northumbria University Student Guide (↖)
Art Direction: Gardiner Richardson

009

BERNAUER STRAßE

PRENZLAUER BERG

PRENZLAUER ALLEE

TORSTRAßE

ALEXANDERPLATZ

U · KARL-MARX-ALLEE

FRIEDRICHSHAIN

UNTER DEN LINDEN

MITTE

LEIPZIGER STRAßE

BERLIN OSTBAHNHOF

WARSCHAUER STRAßE

LATZ

ORANIENSTRAßE

KREUZBERG

SKALITZER STRAßE

ALT-TREPTOW

KOTTBUSSER TOR

HASENHEIDE

HERMANNPLATZ

011

TEMPELHOF

Holiday Destination Maps for Oryx, Qatar Airways Inflight Magazine (↖)
Client & Art Direction: Agency Fish

LLEGE, PORTLAND, OREGON

Map Poster of Reed College, Portland, Oregon for Reed College Student Brochure (↖)
Client & Art Direction: Reed College

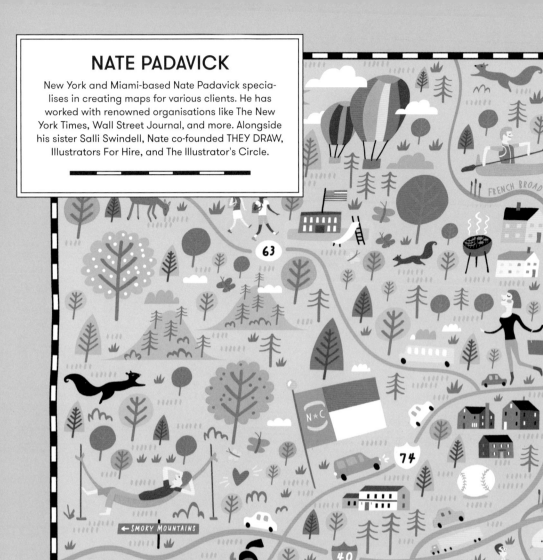

Asheville, North Carolina / Client: True South Puzzle Company ↖

ARCTIC OCEAN

NORTH AMERICA

PACIFIC OCEAN

ATLANTIC OCEAN

SOUTH AMERICA

ANTARCTICA

Sugarloaf, ME

Stowe, VT

Bretton Woods, NH

Lake Placid, NY

Killington, VT

Chautauqua, NY

Seven Springs, PA

Snowshoe, WV

Sugar Mountain, NC

SKI THE USA

Illustrated by Nate Padavick

Ski the USA (↖)
Client: True South Puzzle Company

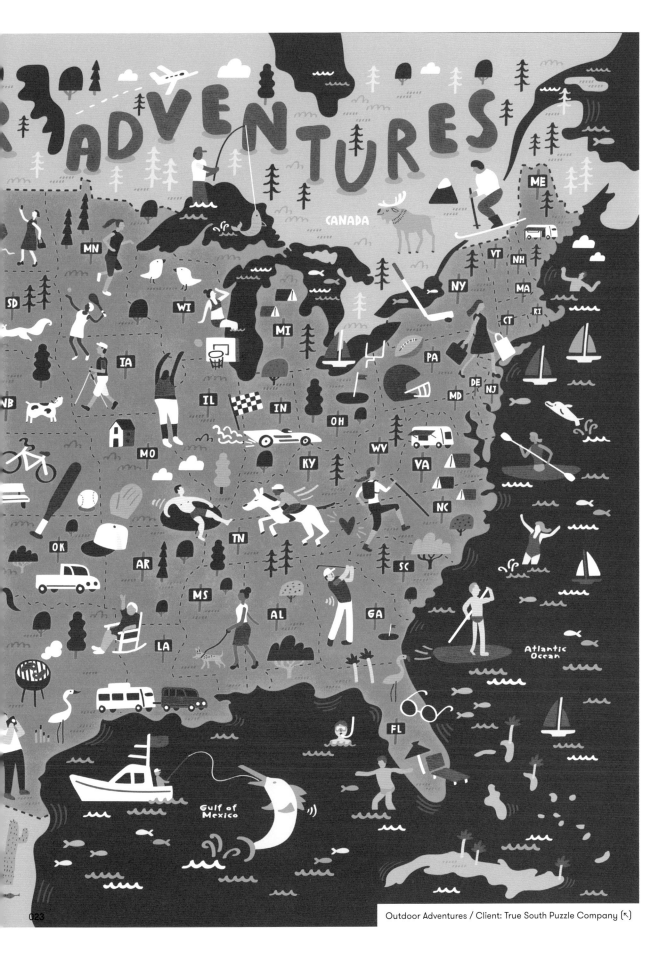

Outdoor Adventures / Client: True South Puzzle Company ↖

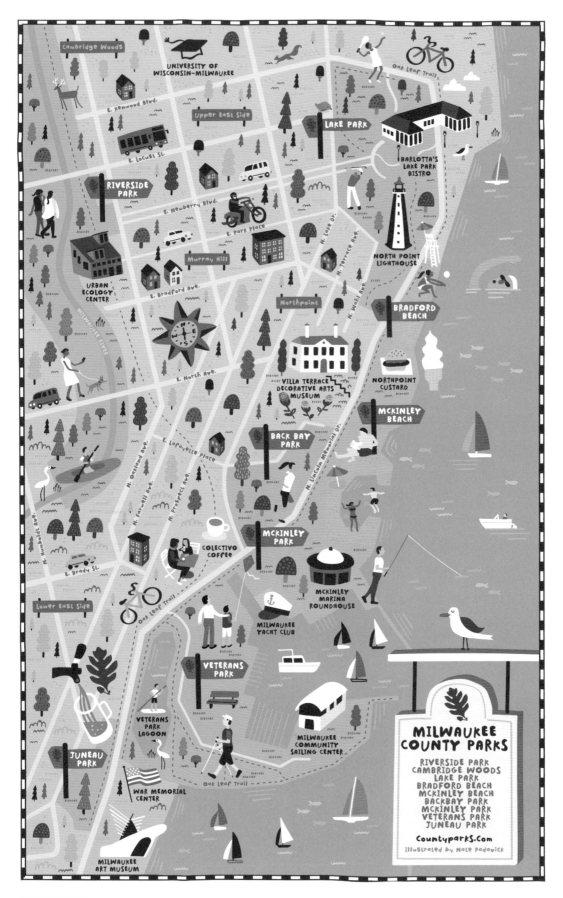

Cambridge Woods

UNIVERSITY OF WISCONSIN–MILWAUKEE

Oak Leaf Trail

E. Kenwood Blvd.

Upper East Side

LAKE PARK

E. Locust St.

RIVERSIDE PARK

BARLOTTA'S LAKE PARK BISTRO

E. Newberry Blvd.

E. Park Place

NORTH POINT LIGHTHOUSE

Murray Hill

URBAN ECOLOGY CENTER

E. Bradford Ave.

Northpoint

BRADFORD BEACH

N. Lake Dr.

N. Terrace Ave.

N. Wahl Ave.

E. North Ave.

VILLA TERRACE DECORATIVE ARTS MUSEUM

NORTHPOINT CUSTARD

MCKINLEY BEACH

BACK BAY PARK

E. Lafayette Place

N. Oakland Ave.

N. Farwell Ave.

N. Prospect Ave.

N. Humboldt Ave.

Oak Leaf Trail

N. Lincoln Memorial Dr.

COLECTIVO COFFEE

MCKINLEY PARK

E. Brady St.

MCKINLEY MARINA ROUNDHOUSE

Lower East Side

Oak Leaf Trail

MILWAUKEE YACHT CLUB

VETERANS PARK

VETERANS PARK LAGOON

MILWAUKEE COMMUNITY SAILING CENTER

JUNEAU PARK

WAR MEMORIAL CENTER

Oak Leaf Trail

MILWAUKEE ART MUSEUM

MILWAUKEE COUNTY PARKS

RIVERSIDE PARK
CAMBRIDGE WOODS
LAKE PARK
BRADFORD BEACH
MCKINLEY BEACH
BACKBAY PARK
MCKINLEY PARK
VETERANS PARK
JUNEAU PARK

CountyparkS.Com

Illustrated by Nate Padavick

Milwaukee Park Banner / Client: Milwaukee County Park Department (←)
Philadelphia, Pennsylvania (↑)

LA RÉUNION

ANTOINE CORBINEAU

Antoine Corbineau is a Nantes-based illustrator who uses a range of mediums and techniques to create beautiful patchworks of texture, imagery, and creative typography. The fun and energetic worlds of his illustrations also communicate simple key messages for the global companies and organisations he works with. He has previously lived in Strasbourg, London, New York and Paris.

SAINT-ANDRÉ

Saint Benoît

PLAINE des PALMISTES

Sainte ROSE

TAMPON

PITON de la Fournaise

SAINT JOSEPH

Saint PHILIPPE

Manapany LES BAINS

A Map of London (↖)

029

A Map of Tokyo (ペ)

031

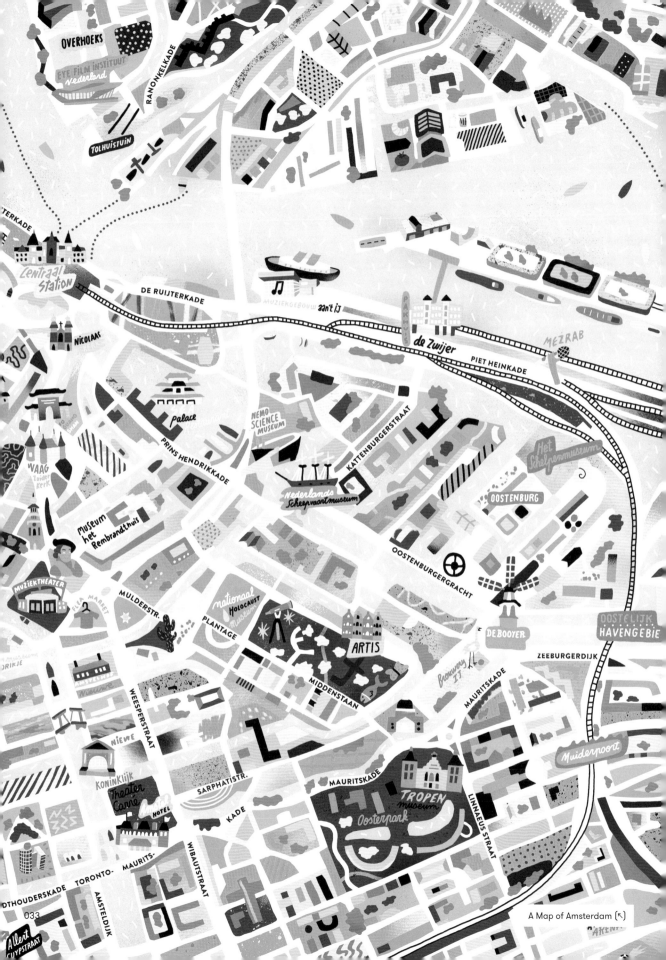

A Map of Amsterdam ↖

033

A Map of Strasbourg

035

A Map of Rennes ↖

037

A Map of Nantes (↖)

ANTOINE CORBINEAU

NEW YORK

Toad Hall

PRINCES HWY

Pambula Town Hall

BRUNKER LN

OREGON ST

The Top Pub

QUONDOLA STREET

PAMBULA Milk Bar VILLAGE

Pambula Village Milk

TOALLA ST

Artessence Gallery

MONARO ST

TOALLA ST

OAKLANDS RD

Toast

Banksia

QUONDOLA STREET

Switchfoot

Vintage TEAROOMS

Cute Shop Vibes

BULLARA STREET

IDLEWILDE CRES

Wild Ryes Baking co.

BAKERY and CAFE   Wild Rye's   COFFEE and ROASTERY

Black Daisy Trading

MERIMBOLA ST

DINGO ST

Longstocking BREWING.

Longstocking Brewery Oaklands

YOWAKA ST

Panboola Wetlands

MUNJE ST

Oaklands

PRINCES HWY

MOUNT DARRAGH RD

PAMBULA RIVER

An illustrated companion to
PAMBULA
Thaua Country

NAT CARROLL

Working from her studio in Djiringanj County, Nat Carroll specialises in illustrations for publications, murals, textiles, and brands. She also has her own line of products and stationery. Her vibrant and whimsical artwork is inspired by the wonders she discovers in her surroundings.

GEORGE ST

WALLAGA LAKE ROAD

Flora & Fauna Reserve

Octopii

LINDO ST

Beardhawk Coffee

777

COFFEE

BEARDHAWK COFFEE

River Rock

WAPENGO ST

MOORHEAD BEACH

WAPENGO ST

BERMAGUI RIVER

Bruce's Ocean Pool

HONEY

E888

Il Passaggio

CAFE
Gelato

bar BAR

Fishermen's Wharf

Shop 7 Gallery

LAMONT STREET

Honorbread

Grower's

MURRAH ST

MURRAH ST

Honorbread

COFFEE   BAKERY

SUMMER
BERMAG

HAY ST

NAT CARROLL

green

Summer in Bermagui ↖

ESTHER GOH

Esther Goh is an illustrator and art director who works with brands across advertising, fashion, publishing, and technology. Her illustrations have been recognised by D&AD, Cannes Design Lions, The One Show, Society of Illustrators, and Asia Tatler, to name a few.

Singapore Botanic Gardens

HOLLAND VILLAGE

ORCHARD ROAD

DEMPSEY HILL

CLARK

TIONG BAHRU

CHINATOW

Mount Faber Park

HARBOURFRONT

SENTOSA

ALYSHA DAWN

Alysha Dawn is a Toronto-based freelance illustrator and designer whose work has appeared internationally in print, large format installations, advertising campaigns, and more. While she focuses on creating unique illustrated maps and commercial work for clients including Adidas, Bissell, Princeton University, Target, Indigo, and more, her work has also been featured in The American Illustration Archive.

CORSO ITALIA
ST. CLAIR AVE. WEST
DUPONT
WALLACE EMERSON
THE ANNEX
ROM
THE JUNCTION
BLOOR
CHRISTIE PITS PARK
BLOORDALE VILLAGE
PALMERSTON LITTLE ITALY
DUFFERIN GROVE
COLLEGE
BROCKTON VILLAGE
AGO
DUNDAS
TRINITY BELLWOODS PARK
QUEEN
CHINATOWN
PARKDALE
KING
HIGH PARK
SUNNYSIDE
NIAGARA
OLD
TORONTO

GREEKTOWN

DANFORTH

DANFORTH

THE BEACHES

LESLIEVILLE

THE DISTILLERY DISTRICT

Tommy THOMPSON PARK

LANDS

LAKE ONTARIO

051

Toronto (2020) (↖)

ade.

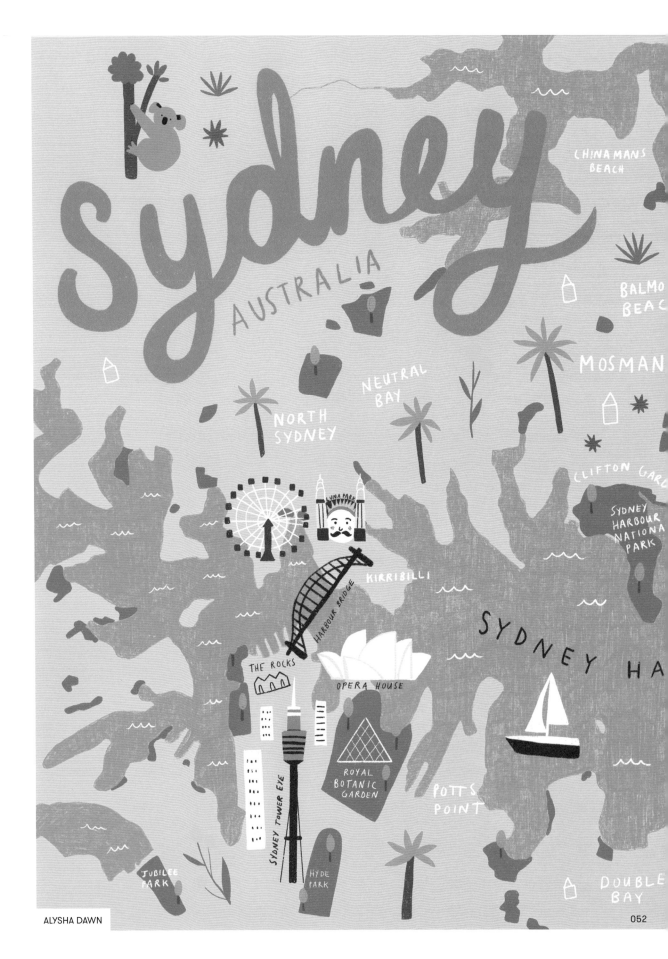

ALYSHA DAWN

052

DOBROYD HEAD

HORNBY
LIGHTHOUSE

WATSONS BAY

GES
TS

R

NIELSEN
PARK

VAUCLUSE

DOVER
HEIGHTS

BONDI BEACH

ROSE BAY

a.d.p.

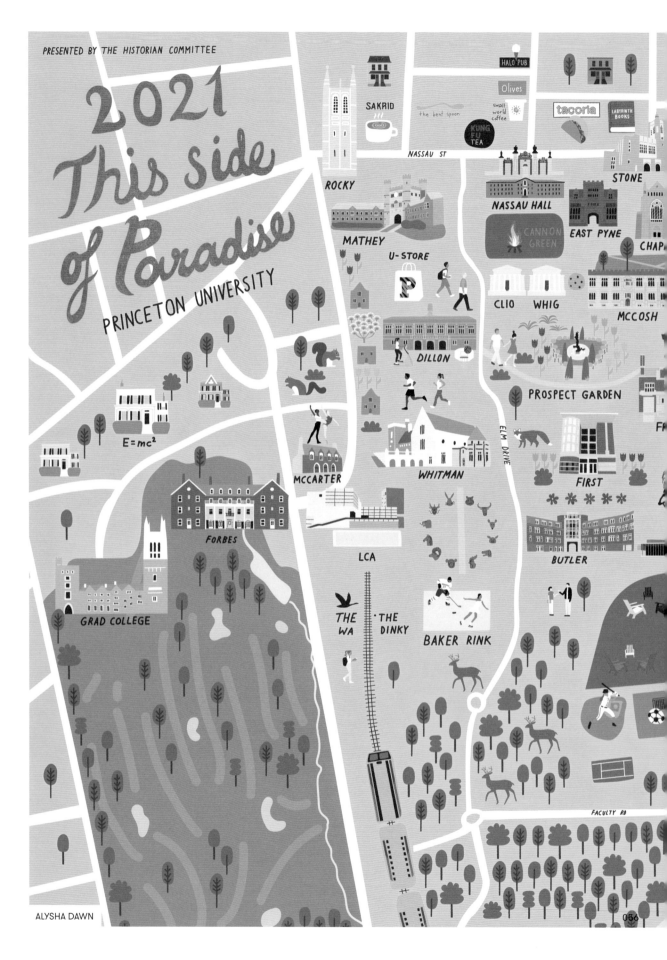

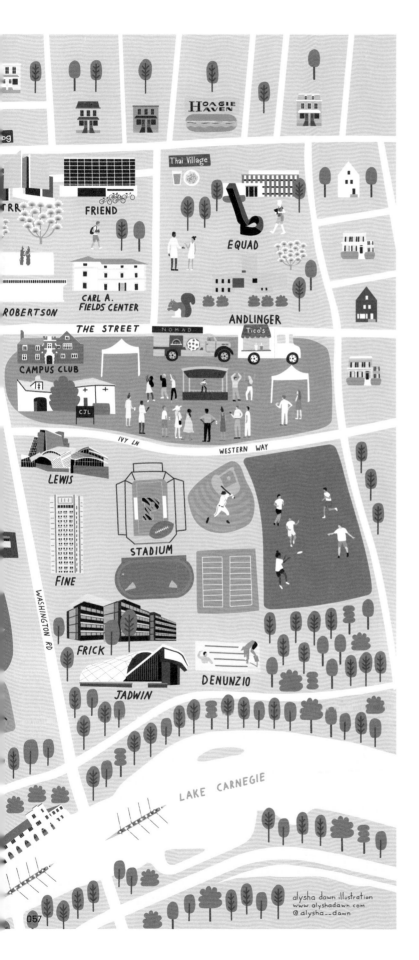

alysha dawn illustration
www.alyshadawn.com
@alysha__dawn

Princeton University (2021) (↖)
Special Credit: Historian Committee

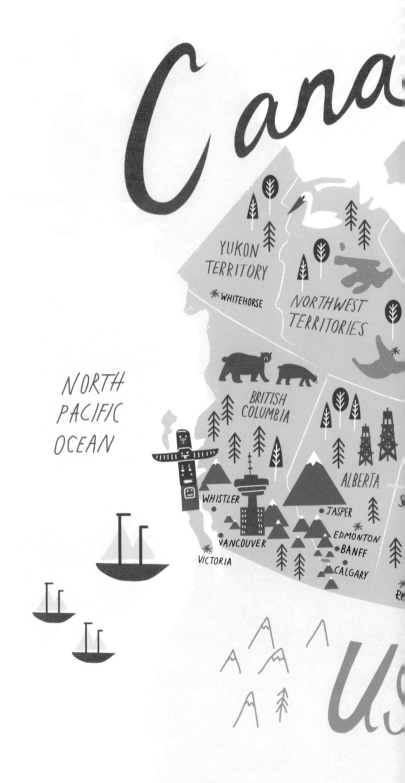

Cana

YUKON
TERRITORY

✳ WHITEHORSE

NORTHWEST
TERRITORIES

NORTH
PACIFIC
OCEAN

BRITISH
COLUMBIA

ALBERTA

WHISTLER

JASPER

VANCOUVER

EDMONTON

VICTORIA

BANFF

CALGARY

US

greenland Sea

IQALUIT

NUNAVUT

HUDSON BAY

MANITOBA

WINNIPEG

ONTARIO

TORNGAT

NEWFOUNDLAND

ST. JOHN'S

QUÉBEC

QUÉBEC CITY

OTTAWA

TORONTO

PEI

CHARLOTTETOWN

NEW BRUNSWICK

FREDERICTON

HALIFAX

NOVA SCOTIA

NORTH ATLANTIC OCEAN

WESTON

AMESBURY

HUMBER
HEIGHTS–
WESTMOUNT

BROOKHAVEN-AMESBURY

BEECHBOROUGH–
GREENBROOK

RICHVIEW

KEELESDALE – EGLINTON
WEST

ROCKCLIFFE
– SMYTHE

PRINCESS–
ROSETHORN

STOCKYARDS
DISTRICT

HUMBER
VALLEY
VILLAGE

THE JUNCTION

HUMBER RIVER

CENTENNIAL PARK

RUNNYMEDE

ISLINGTON

THE
KINGSWAY

HIGH
PARK

SUNNYLEA

SWANSEA

EATONVILLE

MARKLAND
WOOD

YOU ARE HERE!

ISLINGTON – CITY
CENTRE WEST

ETOBICOKE

HUMBER BAY ARCH BRIDGE

Sherway
Gardens

MIMICO

HUMBER
COLLEGE

NEW TORONTO

Lake O

COLONEL
SAM SMITH
PARK

FOREST HILL

RSO ITALIA

ONTO

TRINITY
BELLWOODS

LIBERTY
VILLAGE

SUNNYSIDE BOARD WALK

xio

ALYSHA DAWN
@ alysha__dawn

Illustrated Map of Toronto for Adidas' Sherway Gardens Location (↖)
Agency: Big HQ (2022)

NIKAHO MAP

めくるめく
みずのものがたり
マップ

YUKI MAEDA

Born in 1990, Yuki Maeda lives and worked at a graphic design office in Kyoto, specialising in planning and producing various projects. After about 7 years of experience, Yuki became independent as a freelance illustrator and graphic designer.

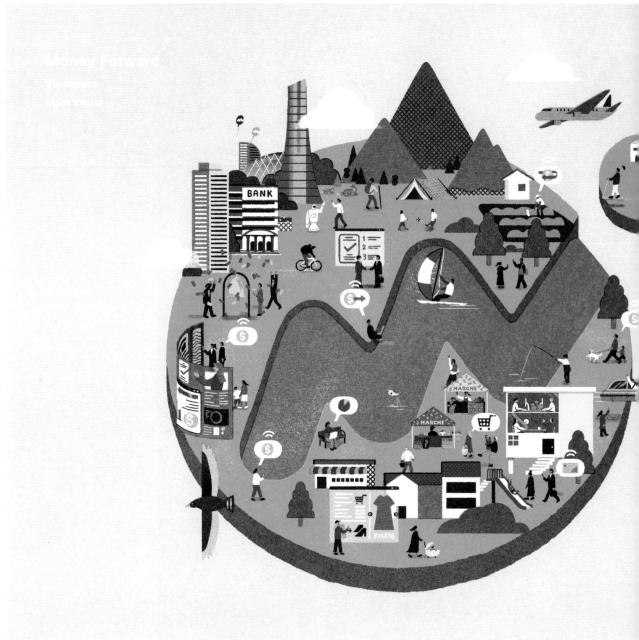

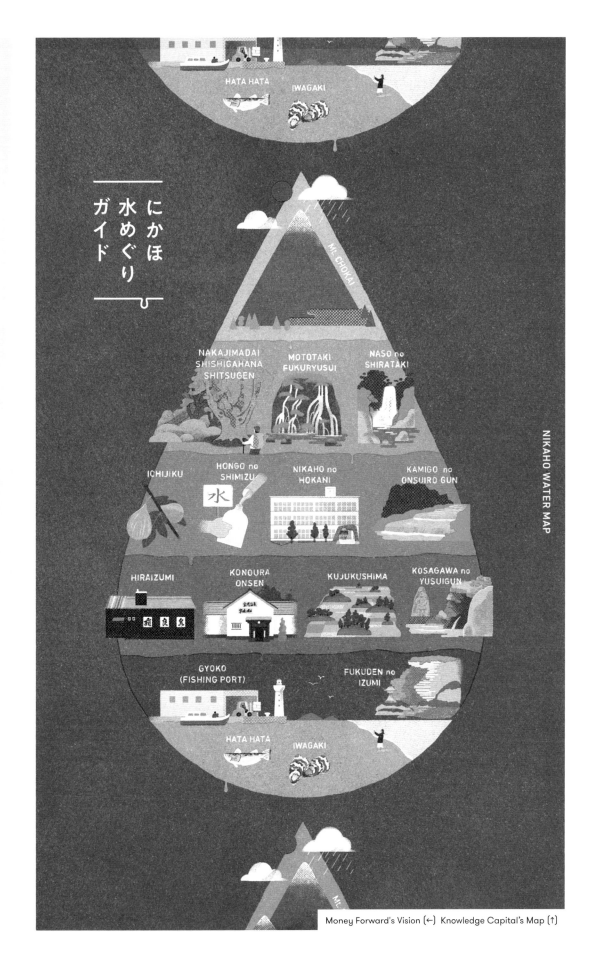

にかほ水めぐりガイド

NIKAHO WATER MAP

ML. CHOKAI

NAKAJIMADAI SHISHIGAHANA SHITSUGEN

MOTOTAKI FUKURYUSUI

NASO no SHIRATAKI

ICHIJIKU

HONGO no SHIMIZU

水

NIKAHO no HOKANI

KAMIGO no ONSUIRO GUN

HIRAIZUMI

KONOURA ONSEN

KUJUKUSHIMA

KOSAGAWA no YUSUIGUN

GYOKO (FISHING PORT)

FUKUDEN no IZUMI

HATA HATA

IWAGAKI

Money Forward's Vision (←) Knowledge Capital's Map (↑)

Knowledge Capital's Map (←)

PARIS
FOOD MAP

BOUILLON PIGALLE

ROSE BA

ARC de TRIOMPHE

La TOUR EIFFEL

SANUKIYA UDON

BE

CRÉMERIE RESTO POLIDOR

Th

SHIBA CAFÉ

POLIDOR

GR du J

GOOD NEWS COFFEE

K-TOWN

ASTRID PRASETIANTI

Astrid Prasetianti is an Indonesian freelance illustrator and graphic designer based in Paris. She loves to experiment with bright colours and playful lines in her work, and is often inspired by her travels, food, and songs she listens to. In her free time, she enjoys taking photos with her film camera, trying new restaurants, and drawing.

LAUREN HUMPHREY

Lauren Humphrey is a London-based illustrator with 10 years of experience. She loves working with bright colours, bold shapes, and is often inspired by plants, animals and anatomy. She enjoys drawing characters and objects with unique proportions and perspectives.

LAUREN HUMPHREY

074

JOYCE FAN

Joyce Fan is based in Portland, Oregon where she is a 3D production artist by day and an illustrator by night. Her illustrative work is often inspired by the small, magical moments in life and oftentimes reminiscent of a child's imagination with all its whimsical adventures.

CITIX60: Portland City Guide (↖)

LAURA SÁSDI

Laura Sásdi is a freelance graphic designer and illustrator based in Budapest who works on book illustrations, maps, exhibition design, and branding. She is inspired by travelling the world and the beauty found in nature.

MONTEGROSSO CINAGLIO

Casa Serra

Fiume Tanaro

ISOLA D'ASTI

Asti

ASTI – TORINO A21

VIGLIANO

CASTELLO DI ANNON

MOMBERCELLI

MONTEGROSSO D'ASTI

AGLIANO TERME

CASTELNUOVO CALCEA

Castello di Costigliole

COSTIGLIOLE D'ASTI

Parco artistico Orme su La Court

MOASCA

S. MARZANO OLIVETO

CASTAGNOLE LANZE

Castello di Moasca

Torre dei Contini

CALOSSO

COAZZOLO

CASTIGLIONE TINELLA

CANELLI

S. STEFANO BELBO

Chiesetta di Coazzolo (Artista: D. Tremlett)

Percorso del Moscato

CASSINA

Tartufo bianco

COSSANO BELBO

LOAZZOLO

Robiola

BUBBIO

ROCCAVERANO ↓

Nocciole

Parco Quir

VESIME ↓   CESSOLE ↓

Parco d'Arte Quarelli

Torr

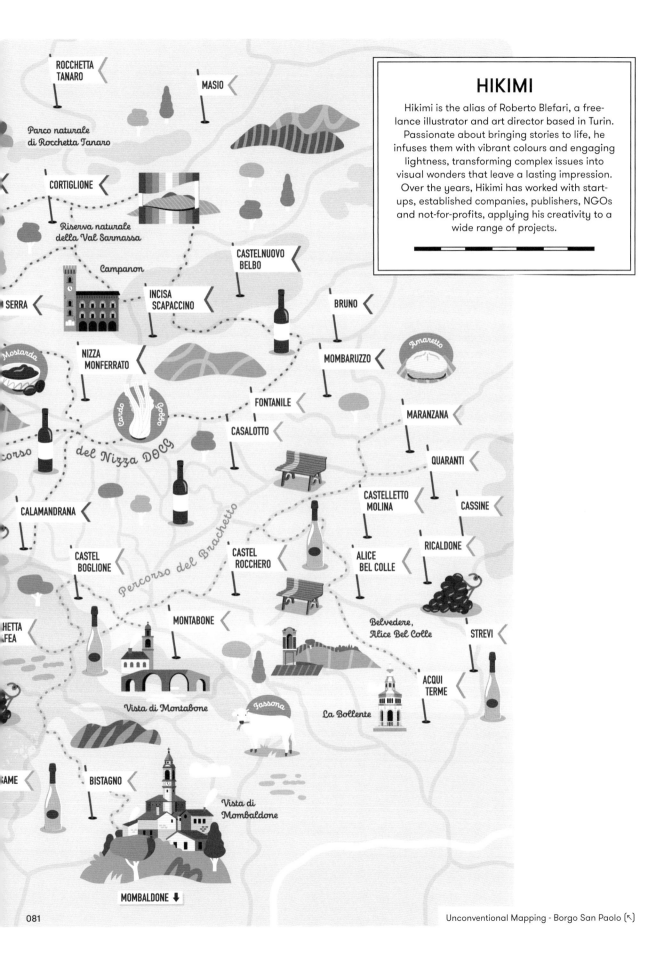

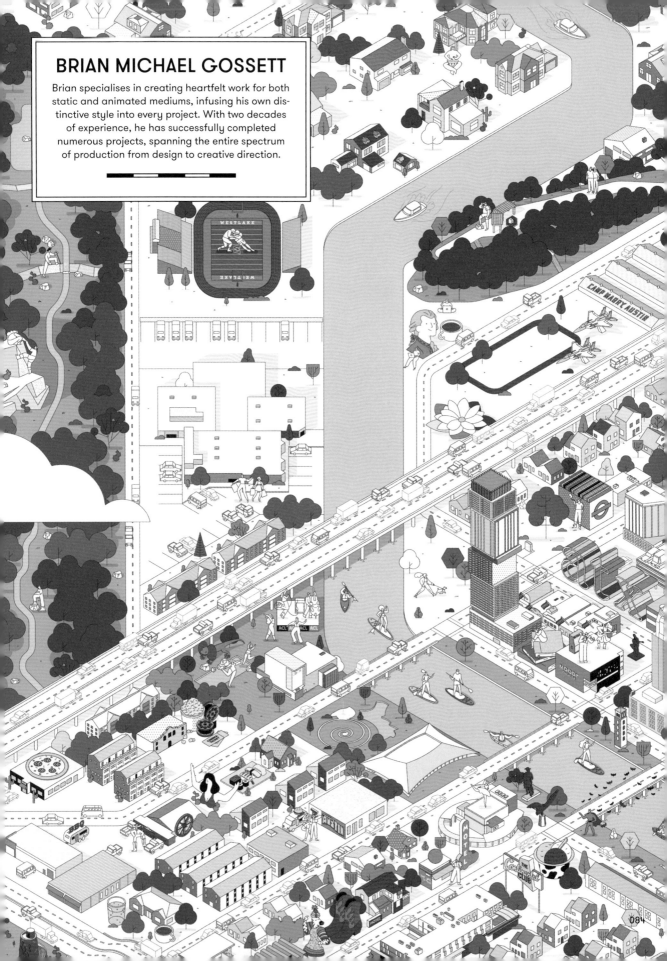

BRIAN MICHAEL GOSSETT

Brian specialises in creating heartfelt work for both static and animated mediums, infusing his own distinctive style into every project. With two decades of experience, he has successfully completed numerous projects, spanning the entire spectrum of production from design to creative direction.

Oracle Map of Austin / Client: Oracle / Art Direction: Pam Caperton (Asterisk Design) (↖)

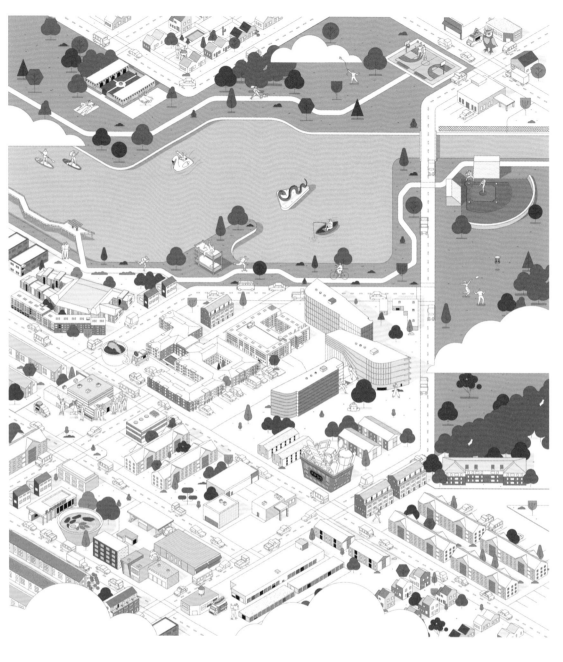

BRIAN MICHAEL GOSSETT

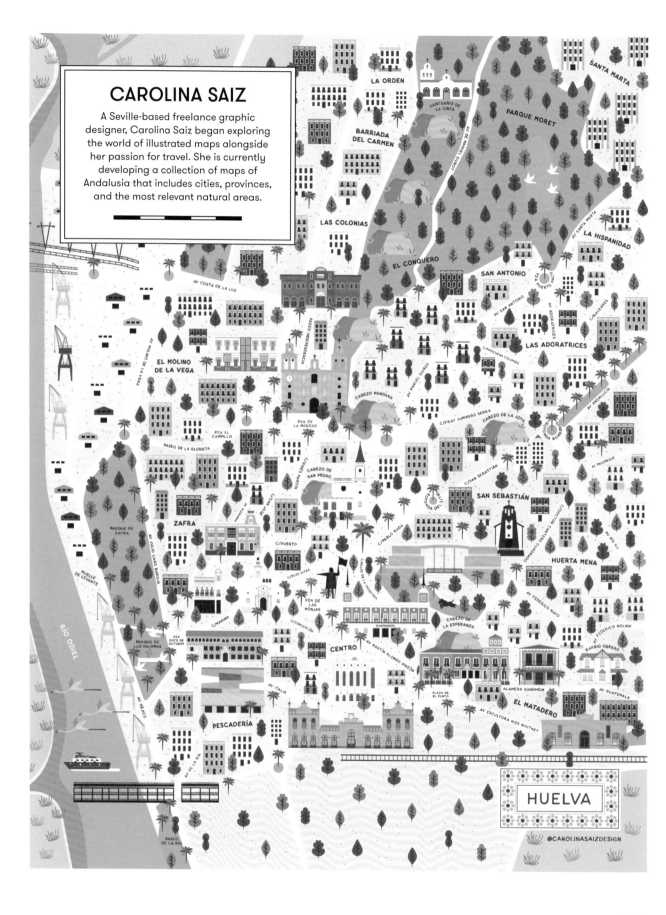

CAROLINA SAIZ

A Seville-based freelance graphic designer, Carolina Saiz began exploring the world of illustrated maps alongside her passion for travel. She is currently developing a collection of maps of Andalusia that includes cities, provinces, and the most relevant natural areas.

HUELVA

@CAROLINASAIZDESIGN

088

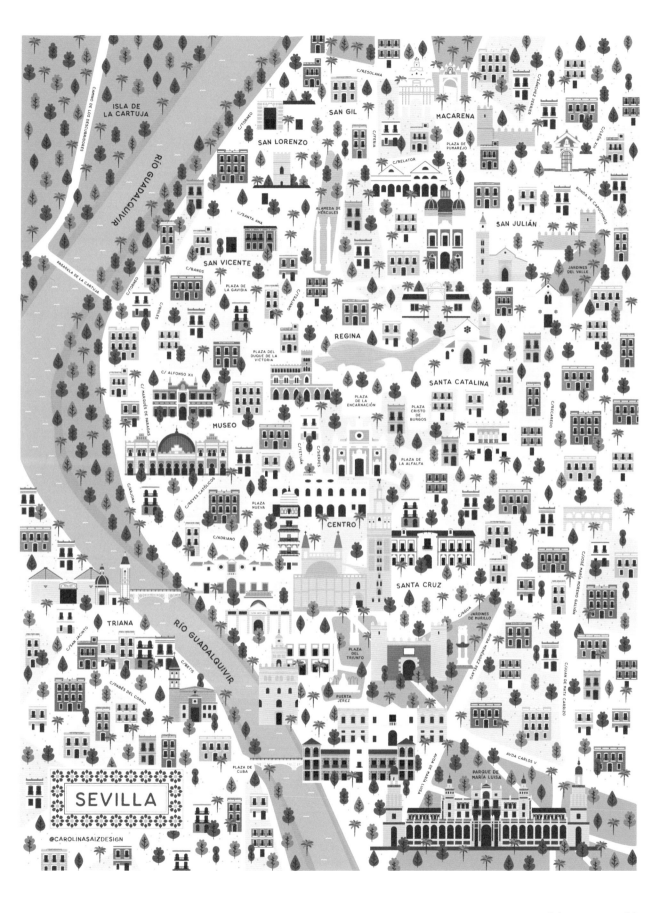

SEVILLA

@CAROLINASAIZDESIGN

Map of Huelva (←) Map of Seville (↑)

ELIZAVETA A. BUINOSOVA

Elizaveta A. Buinosova is an illustrator for children's literature who is also experienced in animation, game design, souvenirs, album covers, and more.

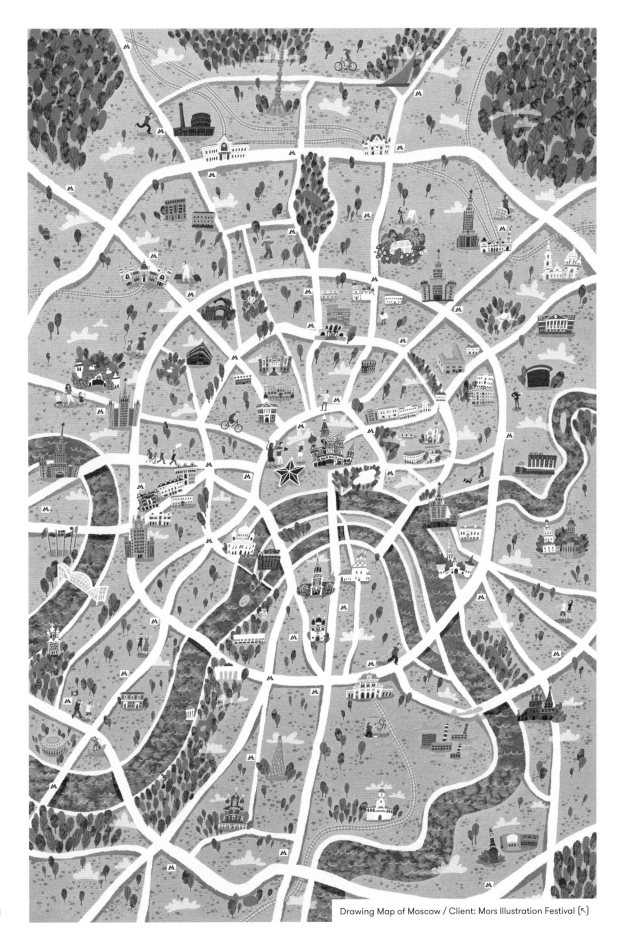

091

Drawing Map of Moscow / Client: Mors Illustration Festival (κ)

FEDERICA UBALDO

Federica is a freelance illustrator and graphic designer who works with publishers, brands, and advertising agencies worldwide. Her playful and quirky style takes inspiration from mid-century visual arts, as well as vintage food advertisements, design from the 1950s, and symphonies.

VIA GALLARATE

SS 11

PARCO DI TRENNO

VIA HARAR

VIALE R SERRA

CORSO SEMPIONE

SAN SIRO

VIALE PISA

VIA SARDEGNA

sante m delle gra

VIALE C DA FORLI'

VIALE E BETZI

VIA F PARRI

SELLA NUOVA

VIA ZURIGO

VIA BISCEGLIE

CAFE'

BAR

VIALE LIGU

ALZAIA

Northbank Map / Client: Northbank Business Improvement District ↖

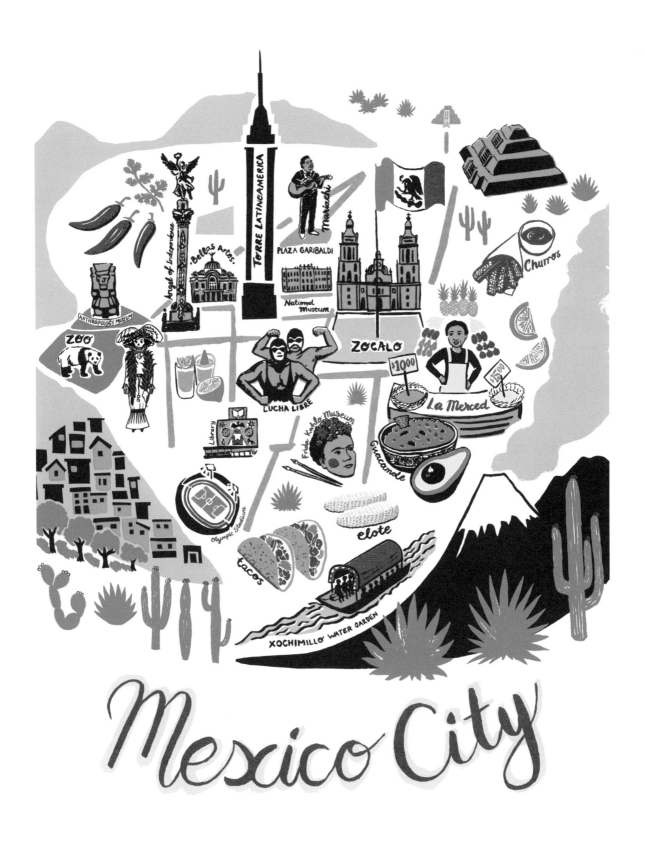

Mexico City

Mexico City (←) Central London (↖)

London's West End Christmas 2020 (↖)
Clients: New West End Company, The Crown Estate, Heart of London
Business Alliance, Shaftesbury Estate, Capco, Portman Estate

Gwinnett County

Lake Lanier

Buford

Tannery Row
Artist Colony

Sugar Hill

Suwanee

Mall of
Georgia

Chattahoochee
River

Infinite
Energy
Center

Suwanee
SculpTour

Coolray
Field

Duluth

Jones
Bridge Park

Southeastern
Railway Museum

85

Peachtree
Corners

Lawrenceville

29

Norcross

Eagle Rock
TV Studios

29

Medieval
Times

Lilburn

124

Aurora
Theatre

BAPS Shri
Swaminarayan
Mandir

78

Snellville

Stone
Mountain
Park

N

W   GC   E

S

Atlanta

Gwinnett County / Client: Atlanta Magazine (↖)

Olivia Brotheridge

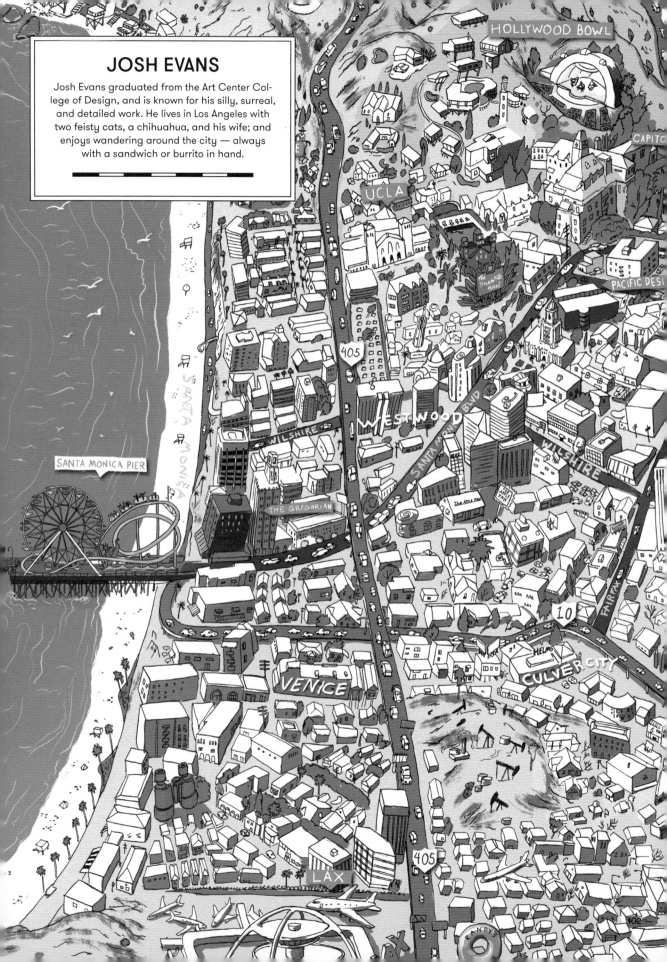

Donaukiez Neighbourhood Map (↗)
Client: Quartiersmanagement Donaustrasse-Nord, Berlin

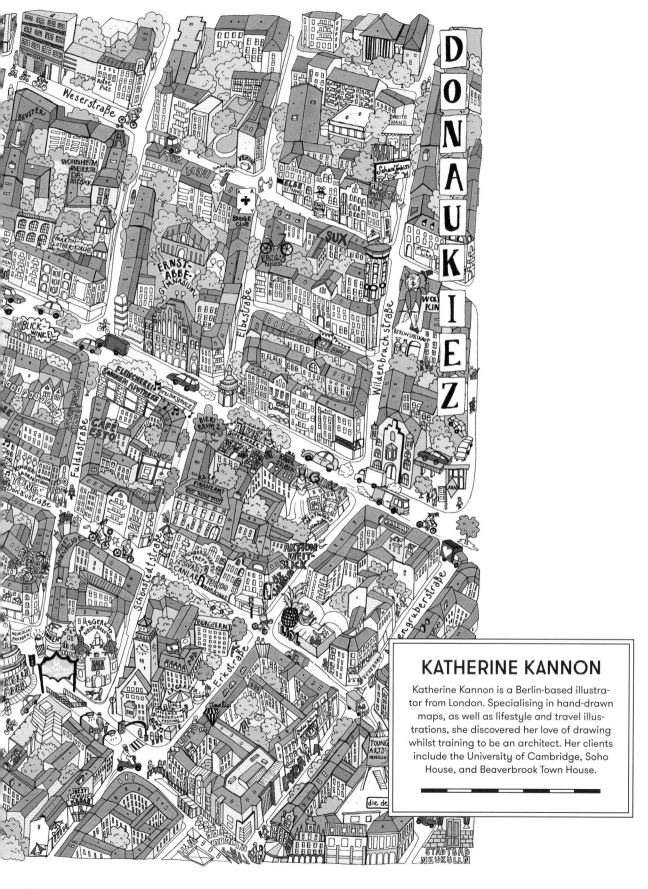

KATHERINE KANNON

Katherine Kannon is a Berlin-based illustrator from London. Specialising in hand-drawn maps, as well as lifestyle and travel illustrations, she discovered her love of drawing whilst training to be an architect. Her clients include the University of Cambridge, Soho House, and Beaverbrook Town House.

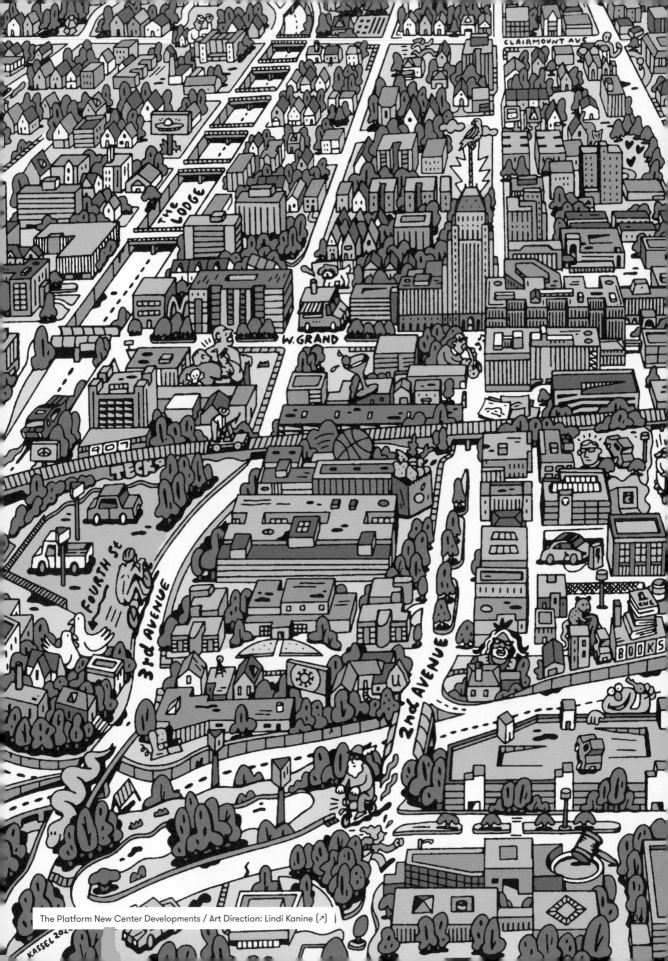

The Platform New Center Developments / Art Direction: Lindi Kanine [↗]

JESSE KASSEL

Jesse Kassel is an illustrator based in Detroit, Michigan. He has been fascinated by maps since childhood and has been illustrating them professionally since 2017.

TAKEaWALK.in Kreuzberg Guide / Art Direction: Thomas Janso (↖)

Taipei Map (↗)

TIBIDABO

TORRE DE COLLSEROLA

SANTA MARIA
DE VALLVIDRERA

OBSERVATORI FABRA

CASA ARNUS

TORRE BELLESGUARD

MUSEU DE LA CIENCIA

COSMOCAIXA

MONESTIR DE PEDRALBES

CONVENT VALLDONZELLA

PORTA FINCA MIRALLES

PAVELLONS GÜELL

LA ROTONDA

PAULAU DE PEDRALBES

CASA BATLLÓ

CAMP NOU

ESTACIÓ DE SANTS

HOSPITAL CLINIC

CASA SAVRACH

CASA AMATLLER

FUNDACIÓ TAPIES

MAGORIA

PALAU NACIONAL

CAIXA FÓRUM

CASA DE LA CARITAT

PALAU DE LA VIRREINA

PAVELLÓ MIES
VAN DER ROHE

MAGBA

POBLE ESPANYOL

MERCAT DE LES FLORS

TORRE CALATRAVA

FUNDACIÓ JOAN MIRÓ

MERCAT
SANT ANTONI

LA BOQUERIA

ESTADI OLIMPIC

PALAU SANT JORDI

LICEU

CASTELL DE MONTJUIC

PALAU GÜELL

REIALS
DRASSANES

112

JORDI SANTJOAN CUNÍ

Jordi Santjoan is a Barcelona-based illustrator who established Tottoristudio. With a lifelong passion for art and painting, he loves everything about maps and headed a book project that catalogued all the small regions and elements of Catalonia, which was published 2 years ago and well-received by the public. Jordi also recently produced the book's second edition.

Catalunya Comarca a Comarca (↖)
Special Credit: Cossetània Edicions

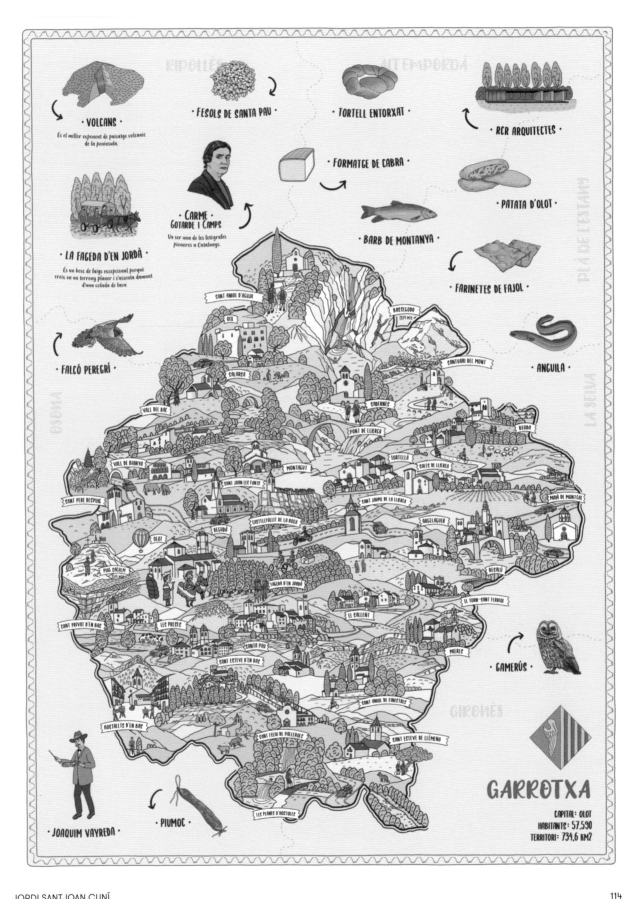

· VOLCANS ·

És el millor exponent de paisatge volcànic de la península.

· FESOLS DE SANTA PAU ·

TORTELL ENTORXAT

· RCR ARQUITECTES ·

FORMATGE DE CABRA

· PATATA D'OLOT ·

· CARME · Gotarde i Camps

Va ser una de les fotògrafes pioneres a Catalunya.

BARB DE MONTANYA

· FARINETES DE FAJOL ·

· LA FAGEDA D'EN JORDÀ ·

És un bosc de faigs excepcional perquè creix en un terreny planer i s'assenta damunt d'una colada de lava

· FALCÓ PEREGRÍ ·

· ANGUILA ·

SANT ANIOL D'AGUJA

BASSEGODA 1373 mts

OIX

SANTUARI DEL MONT

SALARSA

VALL DEL BAC

SADERNES

PONT DE LLIERCA

BEUDA

VALL DE BIANYA

TORTELLÀ

MONTAGUT

SALES DE LLIERCA

SANT JOAN LES FONTS

MANÍ DE MONTCAL

SANT PERE DESPUIG

SANT JAUME DE LA LLIERCA

CASTELLFOLLIT DE LA ROCA

ARGELAGUER

BEGUDÀ

OLOT

PUIG SACALM 1515 mts

FAGEDA D'EN JORDÀ

BESALÚ

EL TORN–SANT FERRIOL

LES PRESES

EL SALLENT

SANT PRIVAT D'EN BAS

SANTA PAU

MIERES

· GAMERÚS ·

SANT ESTEVE D'EN BAS

SANT ANIOL DE FINESTRES

HOSTALETS D'EN BAS

SANT FELIU DE PALLEROLS

SANT ESTEVE DE LLÉMENA

GARROTXA

CAPITAL: OLOT
HABITANTS: 57.530
TERRITORI: 734,6 KM2

· JOAQUIM VAYREDA ·

· PIUMOC ·

LES PLANES D'HOSTOLES

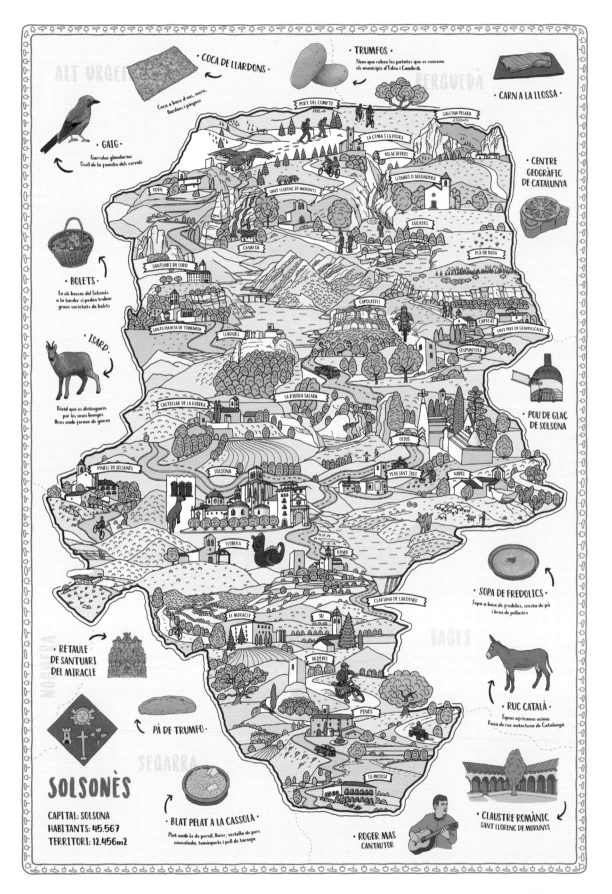

· COCA DE LLARDONS ·

Coca a base d'ous, sucre, llardons i pinyons.

· TRUMFOS ·

Nom que reben les patates que es conreen als municipis d'Oden i Cambrils.

· CARN A LA LLOSSA ·

· GAIG ·

Garrulus glandarius
Ocell de la família dels corvids

· CENTRE GEOGRÀFIC DE CATALUNYA

· BOLETS ·

En els boscos del Solsonès a la tardor si poden trobar grans varietats de bolets

· ISARD ·

Bòvid que es distingueix per les seves banyes llises amb forma de ganxo

· POU DE GLAÇ DE SOLSONA

· SOPA DE FREDOLICS ·

Sopa a base de fredolics, crosta de pà i brou de pollastre

· RETAULE DE SANTUARI DEL MIRACLE

· RUC CATALÀ ·

Equus africanus asinus
Raça de ruc autòctona de Catalunya

· PÀ DE TRUMFO ·

SOLSONÈS

CAPITAL: SOLSONA
HABITANTS: 45.567
TERRITORI: 12.456m2

· BLAT PELAT A LA CASSOLA ·

Plat amb és de pernil, llorer, costella de porc cansalada, tomàquets i pell de taronja

· ROGER MAS
CANTAUTOR

· CLAUSTRE ROMÀNIC
SANT LLORENÇ DE MORUNYS

Map labels: ALT URGELL, BERGUEDÀ, NOGUERA, BAGES, SEGARRA

PORT DEL COMPTE 2392 m, GALLINA PELADA 2320 m, LA COMA I LA PEDRA, VILACIRERES, LLINARS D'AIGUADORA, ODEN, SANT LLORENÇ DE MORUNYS, GUIXERS, CANALDA, PLÀ DE BUSA, SANTUARI DE LORD, CAPOLATELL, CAPOLAT, SANT PERE DE GRAUDESCALES, SANTA EULÀLIA DE TIMONEDA, LLADURS, L'ESPUNYOLA, LA RIBERA SALADA, CASTELLAR DE LA RIBERA, OLIUS, POU DE GLAÇ, PINELL DE SOLSONÈS, SOLSONA, PI DE SANT JUST, NAVES, LLOBERA, RINER, CLARIANA DE CARDENER, EL MIRACLE, SU, ARDÈVOL, PINÓS, LA MOLSORA

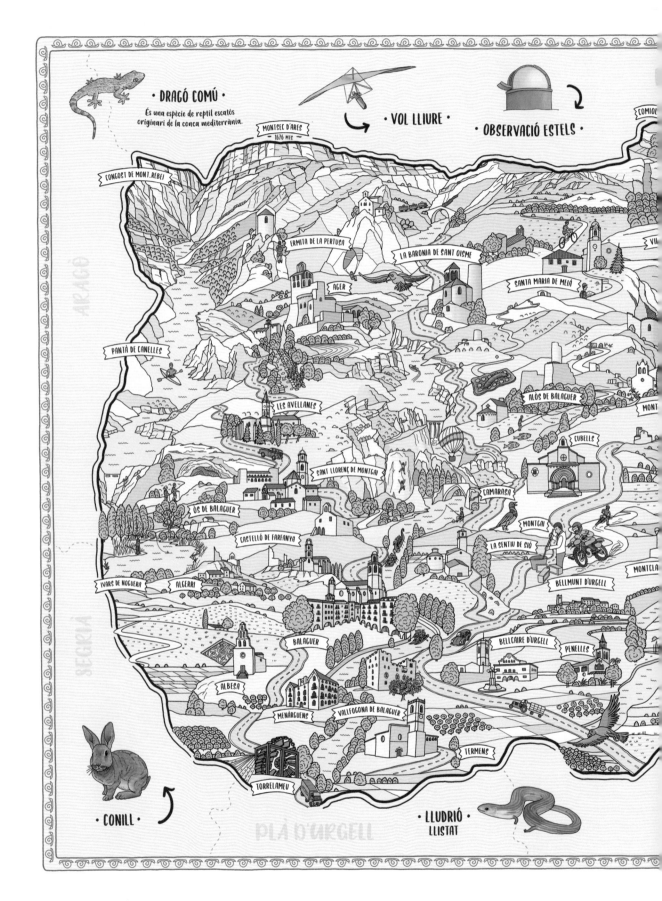

· DRAGÓ COMÚ ·

És una espècie de reptil escatós
originari de la conca mediterrània.

· VOL LLIURE ·

· OBSERVACIÓ ESTELS ·

COMIOL

MONTSEC D'ARES
– 1676 MTS –

CONGOST DE MONT.REBEI

ERMITA DE LA PERTUSA

LA BARONIA DE SANT OISME

ARAGÓ

ÀGER

SANTA MARIA DE MEIÀ

VI

PANTÀ DE CANELLES

ALÒS DE BALAGUER

MONT

LES AVELLANES

CUBELLS

SANT LLORENÇ DE MONTGAI

CAMARASA

ÒS DE BALAGUER

MONTGAI

CASTELLÓ DE FARFANYA

LA SENTIU DE SIÓ

MONTCLA

IVARS DE NOGUERA

ALGERRI

BELLMUNT D'URGELL

SEGRIÀ

BALAGUER

BELLCAIRE D'URGELL

PENELLES

ALBESA

MENÀRGUENS

VALLFOGONA DE BALAGUER

TERMENS

TORRELAMEU

· CONILL ·

PLÀ D'URGELL

· LLUDRIÓ ·

LLISTAT

· COCA DE SAMFAINA ·

Coca salada que porta hortalisses escalivades damunt, típicament pebrera o pebrot, alberginia i ceba.

PALLEROLS

TIURANA

PALAU DE RIALB

GUALTER

VILANOVA DE L'AGUDA

VILVES

ARTESA DE SEGRE

PONTS

OLIOLA

ADADA

CABANABONA

JUSSÀ

ALT URGELL

SOLSONÈS

· PERDIU ·
VERMELLA

· PANADONS ·

Tipus de coques força conegudes durant el segle XVII i XVIII.

SEGARRA

LA NOGUERA

CAPITAL: BALAGUER
HABITANTS: 38.770
TERRITORI: 1.784,1 KM2

· CARGOLS ·
A LA BRUTESCA

URGELL

· CASSOLA DE TROÇ ·

NS

· CASES D'INDIANS ·

· POU DE GLAÇ DE CANYAMARS ·

B

VALLÈS ORIENTAL

LA TRINCA

PÈ

· GUILLA ·

· MADUIXES ·

· COQUETES D'ARENYS ·

· MUSTELA ·

DOSRIUS

SANT VICENÇ DE MONTALT

ORRIUS

VILASSAR DE DALT

ARGENTONA

PREMIÀ DE DALT

CABRERA DE MAR

ALELLA

TEIÀ

CABRILS

MATARÓ

TIANA

VILASSAR DE MAR

PREMIÀ DE MAR

EL MASNOU

MONTGAT

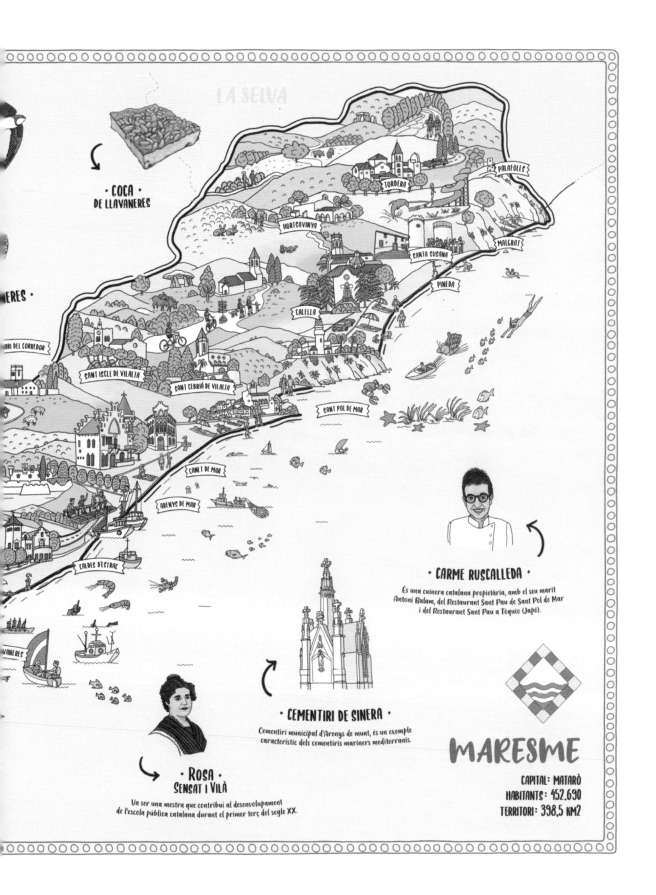

LA SELVA

· COCA ·
DE LLAVANERES

PALAFOLLS

TORDERA

HORTSAVINYA

MALGRAT

SANTA SUSANA

PINEDA

CALELLA

···ERES·

···RI DEL CORREDOR

SANT ISCLE DE VILALTA

SANT CEBRIÀ DE VILALTA

SANT POL DE MAR

CANET DE MAR

ARENYS DE MAR

CALDES D'ESTRAC

···VANERES

· CARME RUSCALLEDA ·

És una cuinera catalana propietària, amb el seu marit
Antoni Balam, del Restaurant Sant Pau de Sant Pol de Mar
i del Restaurant Sant Pau a Tòquio (Japó).

· CEMENTIRI DE SINERA ·

Cementiri municipal d'Arenys de munt, és un exemple
característic dels cementiris mariners mediterranis.

· ROSA ·
SENSAT I VILÀ

Va ser una mestra que contribuí al desenvolupament
de l'escola pública catalana durant el primer terç del segle XX.

MARESME

CAPITAL: MATARÓ
HABITANTS: 452.630
TERRITORI: 398,5 KM2

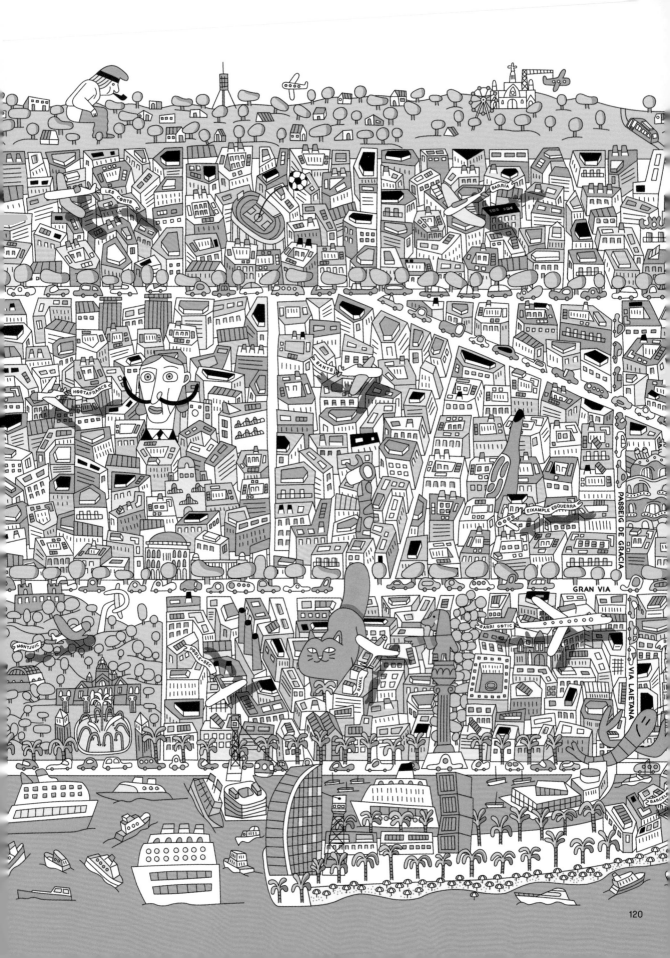

DIEGO MARMOLEJO

Diego Marmolejo is a freelance illustrator based in Barcelona who creates universes made from masses of colour and huggable characters. Childish in appearance but with a mature touch, his illustrations are a strange mixture of minimalism and maximalism.

LOUISE ROSENKRANDS

Danish illustrator Louise Rosenkrands enjoys using bold colours and simple linework to tell stories in images, creating dynamic and cute work that balances a subject's accessibility and comprehensibility with a fun twist.

CITY

Olga Skomorokhova specialises in magazine and book illustrations, as well as commissioned and personal projects. A full-time freelance illustrator in Tallinn, she draws inspiration from travel and people, resulting in the botanical, architectural, and animalistic themes in her art. With a penchant for experimentation and diverse techniques, Olga skilfully captures experiences on canvas through her philosophical mindset, keen observation, and diligence.

London Jigsaw Puzzle (1000 pieces) for Click Europe Ltd. 2021 (↖)

OLGA SKOMOROKHOVA

Rome Jigsaw Puzzle (1000 pieces) for Ningbo Mideer Toys Co Ltd. 2022 ↖

OSTFRIESISCHE INSELN

JUIST

MEMMERT

BORKUM

NORDERNEY

BALTRUM

LANGEOOG

Dornum

Bensers

Norden

KRUMMHÖRN

Suurhusen

Aurich

EMS-JADE-KANAL

Emden

Dollart

Leer

WEENER

Niederlande

Ems

MARTIN HAAKE

W N E S

GRÜNKOH

128

Map of Northern Germany (↖)
Client: daheim Magazine

DEUTSCHLAND

TSCHECHISCHE REPUBLIK

March

Regensburg   Walhalla

Weltenburger
Enge

Passau

Ulm

Donaueschingen

Schlögener
Schlinge

Donau   Bratislava

Wien

ÖSTERREICH

Schloss
Neuschwanstein

SCHWEIZ

SLOWENIEN

Drau

Zagreb

Save

Walzerkönig
Johann Strauss

Venedig

Mailand

ITALIEN

BOSNIEN UND
HERZEGOWINA

Pisa
Florenz

Grosseto

Rom

Book Illustration Map of the River Danubia / Client: Gerstenberg Verlag ↖

Queen Mary's Gardens

Regent's Park

PARK RD

ALLSOP PL

York Gate

YORK GATE

DAUNT BOOKS

DIPTYQUE

St Marylebone Parish Church

FISHWORKS

FISHMONGER & SEAFOOD RESTAURANT

MARYLEBONE RD

YORK ST

GLOUCESTER PLACE

BAKER ST

PADDINGTON ST

DAUNT BOOKS
FISHWORKS
La Fromagerie

MONTAGU PLACE

MARYLEBONE

NEW CAVENDISH

The Wallace Collection

ORCHARD ST

WIGMORE ST

Selfridges

Selfridge

OXFORD ST

Marble Arch

N AUDLEY ST

PARK ST

DUKE ST

BROOK S

SPEAKERS' CORNER

Park

PARK LANE

Grosvenor Square

GROSVENOR ST

MARTIN HAAKE

132

MARYLEBONE RD

PORTLAND PLACE

WEYMOUTH ST

WIGMORE ST

...FORD ST

Tourist Map of Marylebone (←)
Client: Spotlight Magazine

Wall Street

DUMBO
Water St
Front St
Brooklyn Bridge
Flatbush Ave

BROOKLYN HEIGHTS PROMENADE

East River

AMBROSE

BROOKLYN BRIDGE PARK

Noodle Pudding
Sociale BK

CADMAN PLAZA PARK

Clark St
Tazza
Le Pain Quotidien
Montague St
Brooklyn Historical Society

BROOKLYN HEIGHTS

Borough Hall

NY TAXI

EAST RIVER FERRY

Joralemon St

Henry St
Court St

Apple Store

Colonie
Sahadi's

Elsa
Barneys   Trader Joe's
rag & bone

Atlantic Ave

COLUMBIA STREET WATERFRONT DISTRICT

Columbia St

Amity St
COBBLE HILL PARK

Warren St
Clinton St

BROOKLYN RAG & BONE

Smith St
Dean St
Hoyt St
lululemon

COBBLE HILL

Brooklyn Battery Tunnel

Pure Barre

Clover Club
Bird
Cobble Hill Cinemas
Union Market

BIRD

Battersby

Brooklyn Farmacy & Soda Fountain
Barely Disfigured

Bar Bruno
Mazzola Bakery

Van Brunt St

MUS C

President St

Carroll St
Lucali

Sprout San Francisco
Kidville
East One Coffee Roasters
Carroll Park

Union St

Nightingale Nine
Brooklyn Social
Smith Canteen
Wilma Jean

CARROLL GARDENS

Frankies Spuntino
Prime Meats

3rd St

5th St

Buttermilk Channel

MARTIN HAAKE

Brooklyn Neighbourhood Map / Client: Alan Hill Design, NYC (←)
A Guide to Kingston Road Village / Client: BLOK Design (↑)
Abbot Kinney Neighbourhood Map / Client: M Magazine (Le Monde) (↓)

MARTIN HAAKE

136

START HERE

HILLCREST RD

LOEWS HOLLYWOOD HOTEL

N LAS PALMAS AVE

N ORANGE DR

DOLBY THEATRE

HOLLYWOOD AND HIGHLAND CENTER

MADAME TUSSAUDS HOLLYWOOD

TCL CHINESE THEATRE

MUSSO FRANK

THE SPARE ROOM/ THE HOLLYWOOD ROOSEVELT HOTEL

THE HOLLYWOOD MUSEUM

N MCCADDEN PL

SELM

N HIGHLAND AVE

Star Li

IN-N-OUT BURGER

N LA BREA AVE

FRANKLIN AVE

GRACE AVE

N HUDSON AVE

ARGYLE AVE

KIMPTON EVERLY HOTEL

PANTAGES

NO VACANCY

BLACK RABBIT ROSE

LONO

PANTAGES THEATRE

HOLLYWOOD BLVD

D ME

HOLLYWOOD WALK OF FAME

SCHRADER BLVD

CAHUENGA BLVD

W HOLLYWOOD

GOOD TIMES AT DAVEY WAYNE'S

N GOWER ST

THE HIGHLIGHT ROOM/ DREAM HOLLYWOOD

MAMA SHELTER

SEEING

VINE ST

PALEY.

SUNSET BLVD

ARCLIGHT HOLLYWOOD

AMOEBA MUSIC

WILCOX AVE

END HERE

PACIFIC'S CINERAMA THEATRE

MUS C

ЩA

DE LONGPRE AVE

Hollywood Map / Client: Los Angeles Tourism & Convention Board (↖)

MARTIN HAAKE

Victoria Falls

Map of Upper Bavaria / Client: daheim Magazine (←)
Map of the Victoria Falls Area / Client: Ethiopian Airlines (↑)

Thuringia Map / Client: daheim Magazine (←)
VINOS Spain Map / Client: VINOS (↑)

S HEWITT ST

E 1ST ST

ROSE ST

S GAREY ST

S VIGNES ST

ANGEL CITY

216

START HERE

ANGEL CITY
BREWERY

SALT & STRAW

MANUELA

HAUSER & WIRTH

HAMMER
AND SPEAR

BLOOM
SQUARE

E 3RD ST

LUPETTI
PIZZERIA

WURST-
KÜCHE

ALCHEMY WORKS

CAFÉ GRATITUDE

SPACE
MISSION

EIGHTYTWO

THE PIE HOLE

GROUNDWORK

PHILLIP
LIM

AVERY ST

TRACTION AVE

WITTMORE
ARTS
DISTRICT

ARTS DISTRICT
BREWING COMPANY

S SANTE FE AVE

AMAZE-
BOWLS

MERRICK ST

E 4TH ST

S ALAMEDA ST

ARTS DISTRICT CO-OP

123

RESIDENT

MATEO ST

ARTS DISTRICT CO-OP

URTH CAFFÉ

E 5TH ST

COLYTON ST

DOWNTOWN
LOS ANGELES
ARTS
DISTRICT

Los Angeles Arts District Map / Client: Los Angeles Tourism & Convention Board (←)
Map of the Upper East Side in Manhattan / Client: IT Audio, NYC (↑)
Fontenay Neighbourhood Map / Client: The Fontenay Hotel (↓)

JANICE WU

Janice Wu is a Vancouver-based artist and illustrator who works with pencil and gouache on paper. Her illustrations have been featured by the Canada Post, The Walrus, Urban Outfitters, New York Times, The Chinese Canadian Museum, Lancôme, Conde Nast Traveller and more.

149

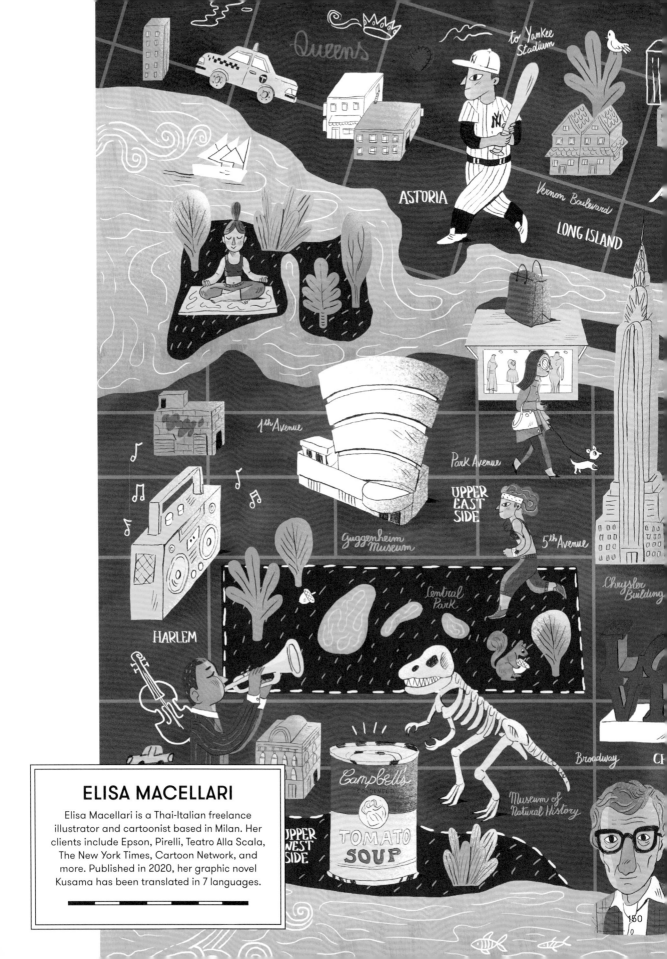

ELISA MACELLARI

Elisa Macellari is a Thai-Italian freelance illustrator and cartoonist based in Milan. Her clients include Epson, Pirelli, Teatro Alla Scala, The New York Times, Cartoon Network, and more. Published in 2020, her graphic novel Kusama has been translated in 7 languages.

150

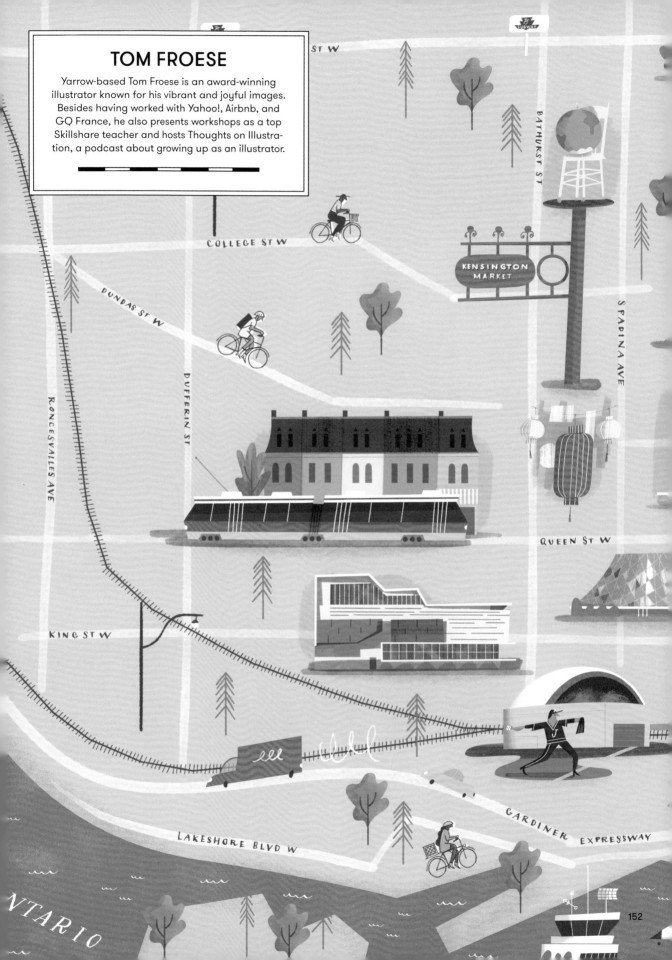

TOM FROESE

Yarrow-based Tom Froese is an award-winning illustrator known for his vibrant and joyful images. Besides having worked with Yahoo!, Airbnb, and GQ France, he also presents workshops as a top Skillshare teacher and hosts Thoughts on Illustration, a podcast about growing up as an illustrator.

ST W

SUBWAY

BATHURST ST

COLLEGE ST W

KENSINGTON MARKET

DUNDAS ST W

S PADINA AVE

DUFFERIN ST

RONCESVALLES AVE

QUEEN ST W

KING ST W

GARDINER EXPRESSWAY

LAKESHORE BLVD W

NTARIO

152

Downtown Toronto / Client: Ryerson University (↖)

PRAGUE

51

DEJVICKÁ

54
53
49
47
48 55 46
56

KORUNOVAČNÍ

63
58
59
6

Letna Beer Garden

69

NÁBŘEŽÍ EDVARDA BENEŠE

Prague Castle

2

6

Charles Bridge

9

50
52

Černy Baby

1
11
13 14
10
8 5

3

Powder Tower

15

44

SVORNOSTI

Petrin T.V. Tower

HOLEČKOVA

41
42

Dancing House

PLZEŇSKÁ

RADLICKÁ

40 43

HOŘEJŠÍ NÁBŘ.

RAŠÍNOVO BŘ.

38

Náplavka

37

Vltava

TOM FROESE

154

DOX Centre for Contemporary Art

LIBEŇSKÝ MOST

Vltava

ARGENTINSKÁ

57

71

72

SOKOLOVSKÁ

NOVOVYSOČANSKÁ

61

65

60

KÁ

ROHANSKÉ NÁBŘ.

80 79 73
78

77

76

74

75

81

34

KONĚVOVA

Wenceslas Square

HUSITSKÁ

TÁBORITSKÁ

27

Žižkov T.V. Tower

30 28

31

VINOHRADSKÁ

29

ŠROBÁROVÁ

23 24

17 25

21

19

16

20

33

32 85

84 86

87

RUSKÁ

22

27

BELGICKÁ

83

82

Havlíček Gardens

88

36

35

NUSELSKÁ

5. KVĚTNÁ

BIMBÓ ÚT

Roman Coast

BIMBÓ ÚT

12

13

Normafa

81

Damube

Margaret Island

MARGIT HÍD

87

7

8 11

9 10

32

21

CSABA U.

82

Matthias Church

ATTILA ÚT

LOVAS ÚT

1

2

Parliament

6

26

20

5

4

ALAGÚT U.

3

Chain Bridge

80

mer's Market

Fisherman's Bastion

Buda Castle

Danube

JAGELLÓ ÚT

HEGYALJA ÚT

ERZSÉBET HÍD

SOMLÓI ÚT

Li

HEGYALJA ÚT

TOM FROESE

VILLÁNYI ÚT

Lady Liberty

Budapest

Chimney Cakes

Budapest Zoo

96

Széchenyi Thermal Baths

94

90

95

91

93

Városliget

92

35

ROTTENBILLER U.

Lángos

KEREPESI ÚT

39

34

36

38

37

St. Stephen's Basilica

40

Heroes' Square

47

ANDRÁSSY ÚT

44

51

49

50

52

55

42 41 43

57

48

46

FIUMEI ÚT

45

KIRÁLY U.

56 54

53

Fröccs

DOB U.

WESSELÉNYI U.

23

Károlyi Garden

31

19

60

33

BRÓDY SÁNDOR U.

59

28

58

18

BAROSS U.

61

47

National Museum

62

63

67

65

78

157

Bálna

69

71

Shop Local: Budapest / Client: Airbnb ↖

70

68

MŰEGYETEM RK.

Danube

est Eye

TOM FROESE

158

Detroit Walking Map (↖)
Client: ICON The Illustration Conference

STANLEY PARK

LOST LAGOON

ENGLISH BAY

W GEORGIA ST

CARDERO ST

COAL HAR

WEST END

BUTE ST

THURLOW ST

BURRARD ST

DAVIE VILLAGE

PACIFIC ST

KITS POINT

CORNWALL AVE

W 1ST

KITSILANO

GRANVILLE ISLAND

FALS

W 4TH

BLANCA ST

ALMA ST

ARBUTUS

CYPRESS ST

BURRARD ST

FAIRVIEW

W 6TH AVE

W BROADWAY

WEST POINT GREY

SOUTH GRANVI

W 12TH AVE

GRANVILLE ST

ARBUTUS RIDGE

DUNBAR SOUTHLANDS

MUSQUEAM FIRST NATION

WESTSIDE

SHAUGHNESSY GRANVILLE

KERRISDALE

33RD AVE

OAK ST

PACIFIC

SOUTH CAMBIE

SW MARINE DRIVE

MARPOLE

TOM FROESE

FRASER RIVE

VANCOUVER HARBOUR

WATERFRONT

W HASTINGS

DUNSMUIR ST

WN

GASTOWN

YALETOWN

ABBOTT ST

CHINATOWN

STATION ST

PRIOR ST

E HASTINGS ST

KEEFER ST

DOWNTOWN
EASTSIDE

HASTINGS SUNRISE

PACIFIC

HWY #1

TERMINAL AVE

E
VAN
ST

CLARK DRIVE

COMMERCIAL DRIVE

GRANDVIEW
WOODLAND

W 2ND

COLUMBIA ST

ONTARIO ST

MAIN ST

W 6TH AVE

E BROADWAY

E 12TH AVE

MOUNT PLEASANT

CRAFT

IPA

FRASER ST

KINGSWAY

RENFREW
HEIGHTS

KENSINGTON
CEDAR
COTTAGE

RILEY
PARK

OAKRIDGE

SUNSET

SOUTH VANCOUVER

VICTORIA
FRASERVIEW

CLAR...

Vancouver Office Mural / Client: Pacific Solutions (↖)

facvltats

Blasco Ibáñez

Carrer de Sant Pius V

Tetuan

CIA

Passeig de l'Albereda

TVRIA

Passeig de la Ciutadella

Carrer de Colom

MIGUEL MONKC

Valencia-based Miguel Monkc is an award-winning advertising and editorial illustrator whose work has been published in renowned newspapers such as The New York Times. He works for advertising agencies and publishers globally to create illustrations, branding, posters, maps, and covers.

The Hood: Chapinero Alto / Art Direction: Christos Hannides (↑) Keller / Art Direction: Lesley Busby (↓)
Client: Hemispheres Magazine / Special Credits: United Airlines, Ink Global

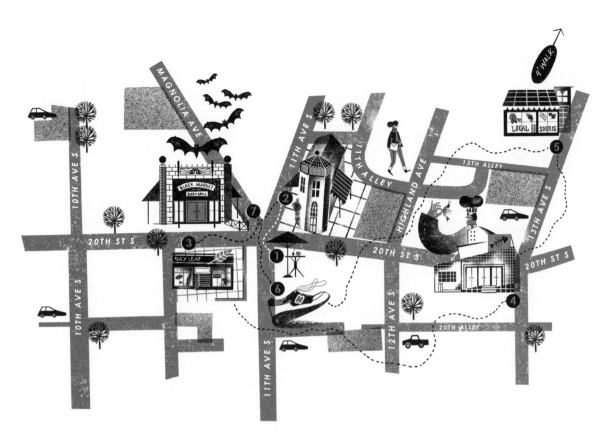

The Hoods: Five Points South / Art Direction: Christos Hannides (↑) Waxahachie / Art Direction: Lesley Busby (↓)
Client: Hemispheres Magazine / Special Credits: United Airlines, Ink Global

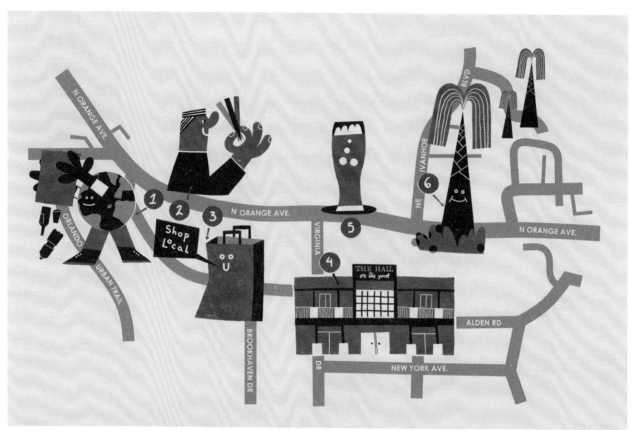

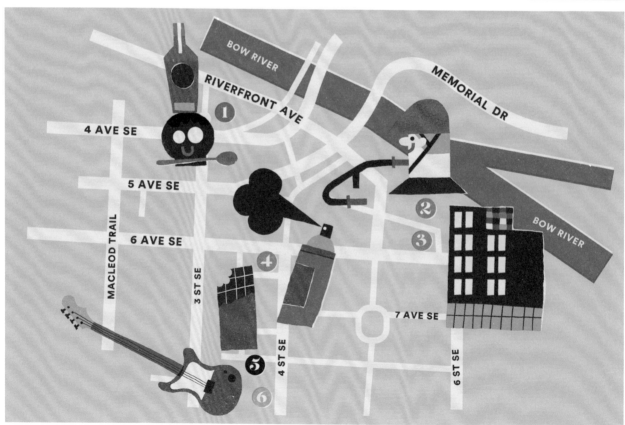

The Hood: Orlando (↑) The Hood: Calgary (↓)
Art Direction: Christos Hannides / Client: Hemispheres Magazine / Special Credits: United Airlines, Ink Global

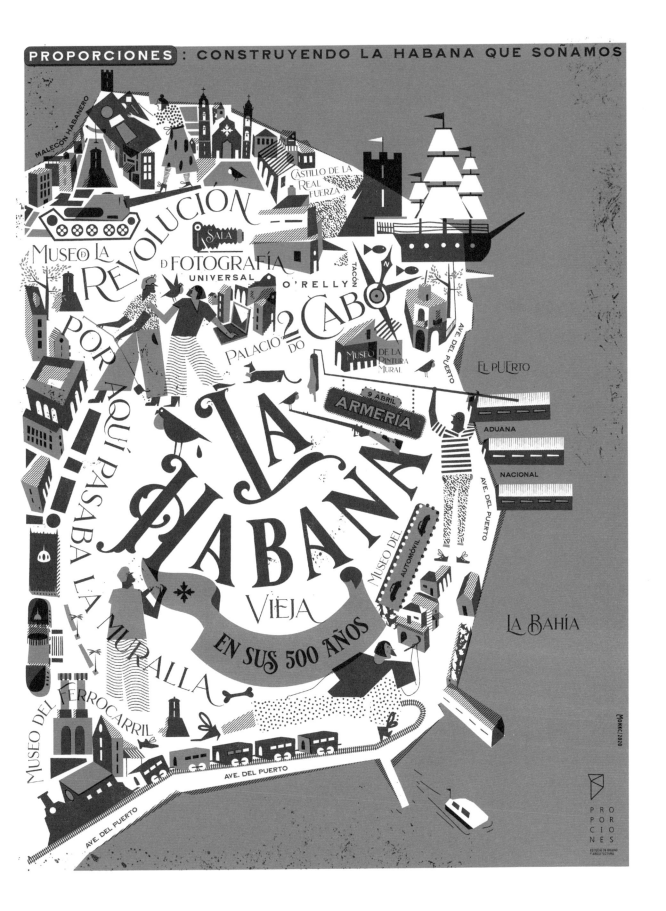

Havana City, 500th Anniversary / Client: Proporciones (↖)

Av Carnot

Arc de Triomphe

Av Friedland

Av Foch

Av Kléber

Av d'Iéna

Av des Champs-É

Av Georges Mandel

Seine

Quai des

Av Paul Doumer

Av Bosquet

Quai de Grenelle

Bd Saint Germain

Tour Eiffel

Musée Rodin

Joffre

METRO

Place

Rue de Sèvres

Bd de Grenelle

Av Émile Zola

Bd Pasteur

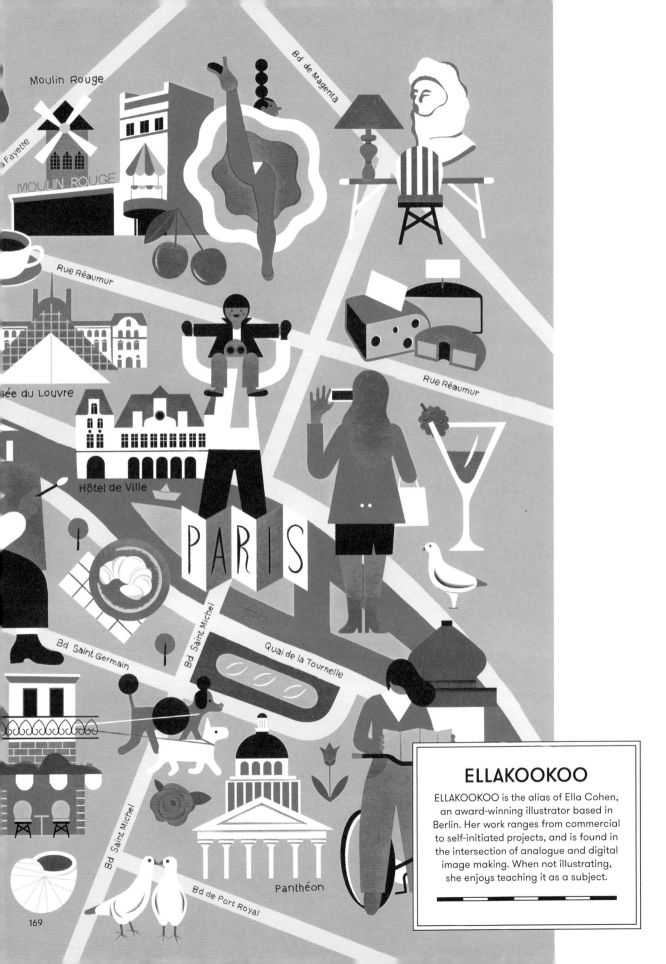

Moulin Rouge

Bd de Magenta

a Fayette

MOULIN ROUGE

Rue Réaumur

Rue Réaumur

sée du Louvre

Hôtel de Ville

PARIS

Bd Saint Germain

Bd Saint Michel

Quai de la Tournelle

Bd Saint Michel

Panthéon

Bd de Port Royal

ELLAKOOKOO

ELLAKOOKOO is the alias of Ella Cohen, an award-winning illustrator based in Berlin. Her work ranges from commercial to self-initiated projects, and is found in the intersection of analogue and digital image making. When not illustrating, she enjoys teaching it as a subject.

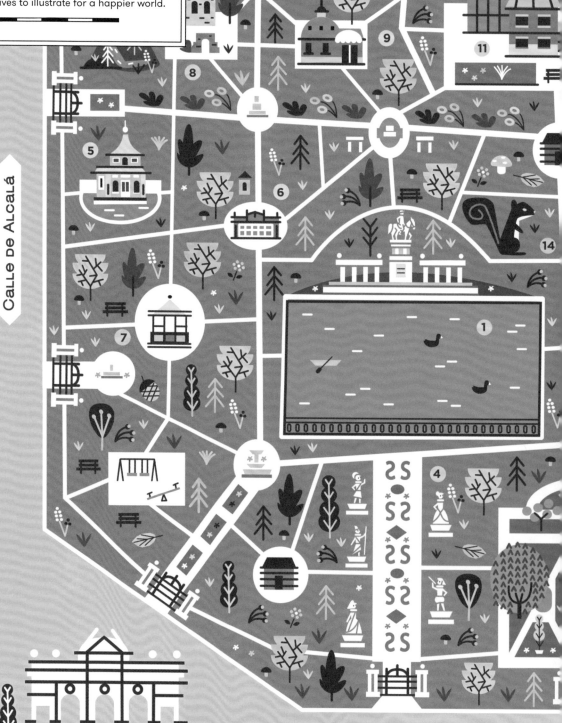

VICTORIA FERNÁNDEZ

Madrid-based Victoria Fernández was previously a copywriter, stop-motion animator, and an on-air digital designer before becoming a freelance illustrator in 2011. Inspired by love, naive art, mid-century modern design, and the wild, she strives to illustrate for a happier world.

Calle de Alcalá

2019

Calle de Alfonso

EL RETIRO

otoño / autumn

superminimaps.com ♥ victoriafernandez.me

MAPAS DEL RETIRO / Design & Editing: Ana Laya Oramas (↖)

N O S E

Avenida

Calle de Alcalá

9

8

15

6

5

1

7

14

4

11

10

2019

VICTORIA FERNÁNDEZ

Calle de Alfonso

EL RETIRO

verano / summer

Calle del Poeta Esteban Villegas

Calle de Alcalá

9

8

5

6

15

1

7

11

4

11

2019

EL RETIRO
primavera / spring

Calle del Poeta Esteban Villegas

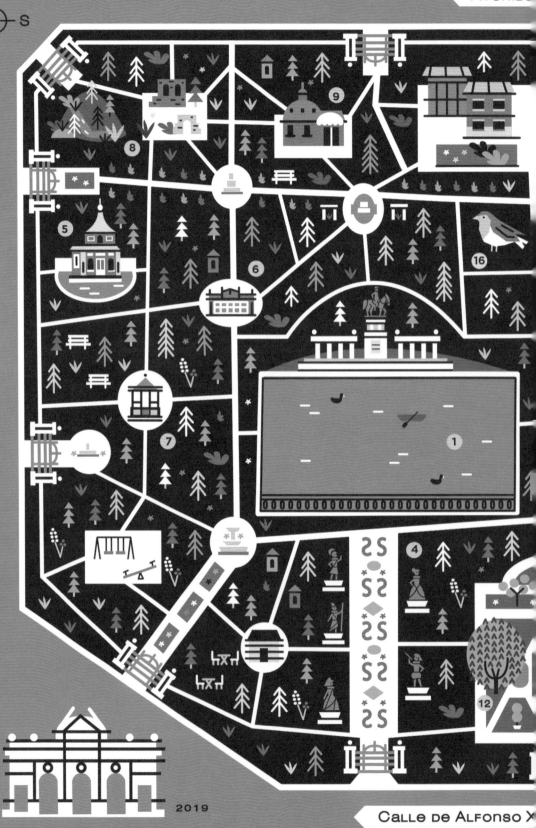

Calle de Alcalá

2019

Calle de Alfonso X

ndez Pelayo

EL RETIRO

invierno / winter

Calle del Poeta Esteban Villegas

supermínimaps.com ♥ victoriafernandez.me

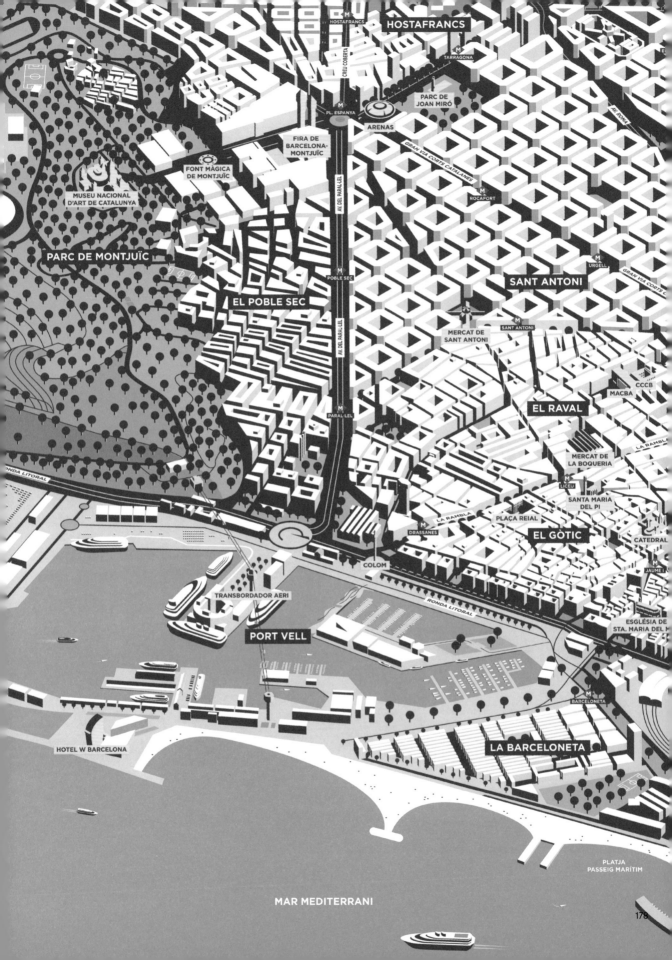

HOSTAFRANCS

HOSTAFRANCS

TARRAGONA

PARC DE
JOAN MIRÓ

PL. ESPANYA

ARENAS

GRAN VIA CORTS CATALANES

AV. ROMA

FIRA DE
BARCELONA-
MONTJUÏC

FONT MÀGICA
DE MONTJUÏC

ROCAFORT

MUSEU NACIONAL
D'ART DE CATALUNYA

URGELL

SANT ANTONI

GRAN VIA CORTS C.

PARC DE MONTJUÏC

EL POBLE SEC

POBLE SEC

MERCAT DE
SANT ANTONI

SANT ANTONI

CCCB

MACBA

EL RAVAL

PARAL·LEL

LA RAMBLA

MERCAT DE
LA BOQUERIA

LICEU

SANTA MARIA
DEL PI

PLAÇA REIAL

EL GÒTIC

CATEDRAL

DRASSANES

LA RAMBLA

JAUME I

COLOM

ESGLÉSIA DE
STA. MARIA DEL M

RONDA LITORAL

RONDA LITORAL

TRANSBORDADOR AERI

PORT VELL

BARCELONETA

HOTEL W BARCELONA

LA BARCELONETA

PLATJA
PASSEIG MARÍTIM

MAR MEDITERRANI

ESCOLA INDUSTRIAL

HOSPITAL CLÍNIC

L'ESQUERRA DE L'EIXAMPLE

AV. DIAGONAL

FORMA

Forma is a graphic design and illustration studio based in Barcelona aiming to assist clients in creating or enhancing their visual identity. Approaching every project with a focus on generating innovative ideas and combining creativity and simplicity, the team delivers designs that leave a lasting impact.

CARRER DE BALMES

RAMBLA DE CATALUNYA

PASSEIG DE GRÀCIA

DIAGONAL

LA PEDRERA

AV. DIAGONAL

JOAN

CASA BATLLÓ

PG. SANT JOAN

PASSEIG DE GRÀCIA

LA DRETA DE L'EIXAMPLE

VERDAGUER

LA SAGRADA FAMÍLIA

PLAÇA DE CATALUNYA

DE CATALUNYA

GRAN VIA CORTS CATALANES

GIRONA

AV. DIAGONAL

LA SAGRADA FAMÍLIA

VIA LAIETANA

URQUINAONA

TETUAN

SAGRADA FAMÍLIA

PG. SANT JOAN

GRAN VIA CORTS CATALANES

MONUMENTAL

ARC DE TRIOMF

ARC DE TRIOMF

PLAÇA DE TOROS MONUMENTAL

AV. DIAGONAL

PARC DE LA CIUTADELLA

AV. MERIDIANA

MARINA

TEATRE NACIONAL DE CATALUNYA

GLÒRIES

AV. ME

MUSEU DEL DISSENY DE BARCELONA

LA VILA OLÍMPICA

BOGATELL

TORRE AGBAR

CARRER DE LA MARINA

CIUTADELLA / VILA OLÍMPICA

RRES MAPFRE

LLACUNA

POBLENOU

RONDA LITORAL

PLATJA

GOLDEN GATE NATIONAL RECREATION AREA

MARIN HEADLANDS

GOLDEN GATE BRIDGE

PALACE OF FINE ARTS

THE WALT DISNEY FAMILY MUSEUM

PEACE P.

BAKER BEACH

VETERANS BLVD

LEGION OF HONOR

CALIFORNIA ST

LANDS END

GEARY BLVD

PAINT

DE YOUNG MUSEUM

THE CASTRO

GOLDEN GATE PARK

CONSERVATORY OF FLOWERS

TWIN PEAKS

19TH AVE

CESAR

STERN GROVE

ANDREA NGUYEN

Andrea Nguyen is a Creative Design Lead at Airbnb, where she has worked since 2013. In her spare time, she loves to illustrate and cook. She currently resides in Orange County with her husband, daughter, and bunny.

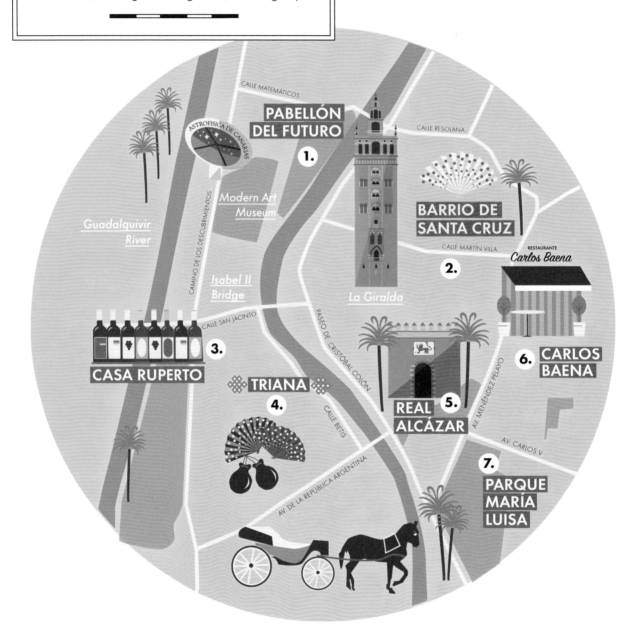

ANA LAYA ORAMAS

Ana is a journalist who works as a freelance creative director, brand strategist, and trend researcher. She is also the creator of two projects: The Procrastinator Times and Superminimaps. Ana loves telling stories, whether it's designing creative campaigns or experiences, writing articles and newsletters, shooting and editing videos, or making maps.

CALLE MATEMÁTICOS

PABELLÓN DEL FUTURO
1.

CALLE RESOLANA

ASTROFÍSICA DE CANARIAS

Modern Art Museum

BARRIO DE SANTA CRUZ

Guadalquivir River

CALLE MARTÍN VILLA

RESTAURANTE
Carlos Baena

Isabel II Bridge

CAMINO DE LOS DESCUBRIMIENTOS

La Giralda

2.

CALLE SAN JACINTO

PASEO DE CRISTÓBAL COLÓN

CASA RUPERTO
3.

CARLOS BAENA
6.

TRIANA
4.

REAL ALCÁZAR
5.

AV. MENÉNDEZ PELAYO

CALLE BETIS

AV. CARLOS V

AV. DE LA REPÚBLICA ARGENTINA

PARQUE MARÍA LUISA
7.

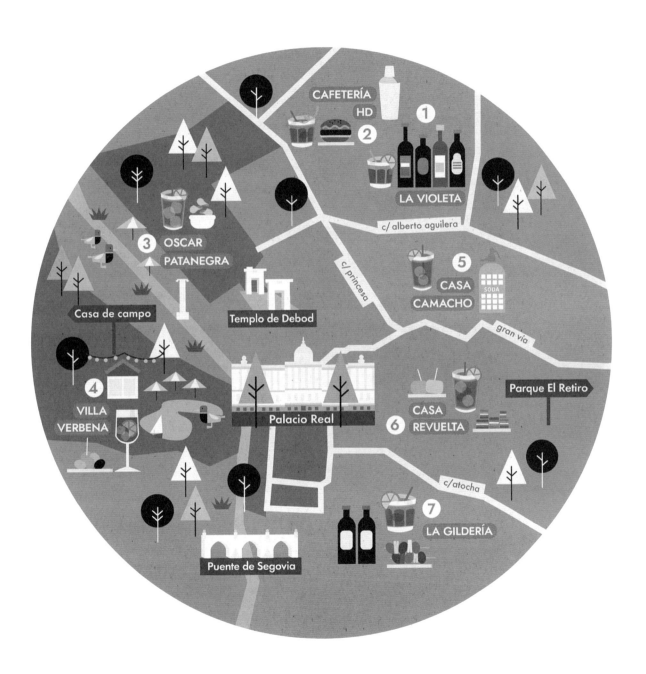

Đồng Xuân Market

84 Hàng Đào

Đường Trần Quang Khải

Hàng Bồ

Hàng Bông

Hoàn Kiếm Lake

7 LE BOUT
DU MONDE

5 JARDINS SUSPENDUS

1 LIRE À LA PLAGE

LA PETITE RADE

6

église
saint-joseph

un été
au havre

perret

le volcan

3 SALON DES
NAVIGATEURS

LE HAVRE PORTE DE L'EUROPE

4 LA HALLE
AUX POISSONS

MUSEÉ D'ART MODERNE
ANDRÉ MALRAUX

185

Let's Eat It All Hanoi Minimap / Design & Editing: Ana Laya Oramas / Illustration: Victoria Fernández (←)
Le Havre Summertime Minimap (↑)

Morocco

Western
Sahara

Tunisia

ALGERIA

LIB

Mauri-
tania

MALI

Niger

Senegal

Gambia

Guinea

Sierra
Leone

Liberia

Côte
d'Ivoire

Burkina
Faso

Ghana

Togo

Benin

Nigeria

Cha

Cameroon

Central

Equatorial
Guinea

Gabon

Republic
of
the
Congo

Atlantic
Ocean

ANGO

Namibi

Africa

TRISTA YEN

Trista Yen is an illustrator and children's book author who graduated from the University of Brighton in sequential design and illustration. With a passion for education, psychology, and mental health, her art often features animals, particularly bears. She mainly works with children's book publishers, but is also expanding into greeting cards and surface design. In 2021, she launched her brand Miss Noproblem, applying her designs to various products.

Egypt

SUDAN

Eritrea

Djibouti

ETHIOPIA

South Sudan

SOMALIA

Uganda

KENYA

Rwanda

Burundi

Tanzania

Indian Ocean

ZAMBIA

MALAWI

Madagascar

Zimbabwe

NA

Eswatini

Lesotho

N

W

E

S

Arctic Ocean

Labrador Sea

Pacific
Ocean

Hudson
Bay

CANADA

UNITED STATES

Atlantic
Ocean

BAHAMAS

Dominican
Republic

CUBA   Haiti

Jamaica

Caribbean
Sea

Belize

Guate-
mala   Honduras

El Salvador   Nica-
ragua

COSTA RICA   PANAMA

North
America

N
W   E
S

Colombia

Ecuador

PERU

Guyana

Suri-
name

French
Guiana

Bolivia

Pacific
Ocean

Paraguay

CHILE

Uruguay

Atlantic
Ocean

Argentina

South
America

N
W E
S

EuRope

Iceland
Russia
Norway
Sweden
Estonia
Latvia
Lithuania
North Sea
Denmark
United Kingdom
Ireland
Netherlands
Belgium
Luxembourg
Germany
Czech Replic
Poland
Slovakia
Ukraine
Switzerland
Austria
Slovenia
Hungary
Romania
Moldova
Croatia
Bosnia & Herzegovina
Serbia
Montenegro
Kosovo
Bulgaria
Macedonia
Atlantic Ocean
Portugal
Spain
Greece
Turkey
Tyrrhenian Sea
Mediterranean Sea

N
W E
S

The British Isles

Shetland Islands

Outer Hebrides

North Sea

Isle of Skye

Inverness

Aberdeen

Atlantic Ocean

SCOTLAND

Dundee

Glasgow

Edinburgh

Newcastle-upon-Tyne

NORTHEN IRELAND

Belfast

Isle of Man

York

ENGLAND

Dublin

Liverpool

Nottingham

IRELAND

Irish Sea

Birmingham

Norwich

WALES

Oxford

Cardiff

Bristol

Bath

London

Dover

Celtic Sea

Brighton

BRIGHTON PIER

Plymouth

N
W E
S

English Channel

Europe (←) Isle of Britain (↖)

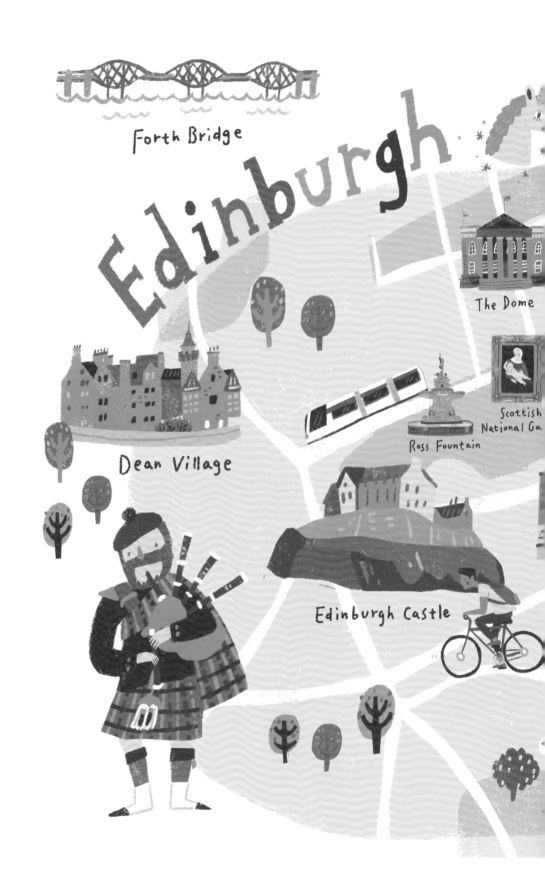

Forth Bridge

Edinburgh

The Dome

Dean Village

Ross Fountain

Scottish
National Ga

Edinburgh Castle

Royal Yacht
Britannia

Calton Hill

Monument

Canongate kirk

Royal Mile

Holyrood House

Street

reyfriars Bobby

Arthur's Seat

eadows

VICKY SCOTT

A freelance illustrator based in Sheffield, Vicky Scott's nostalgic and quirky work is inspired by nature, travel, Art Deco architecture, and 60s psychedelic posters. Employing various techniques like drawing, paper collaging, painting, and Photoshop, she has collaborated with a diverse range of clients such as Microsoft, Cheltenham Festivals, and VW. Her recent map illustrations combine her love for exploring new places and highlighting captivating details.

London

Regents Park

Diana Memorial Fountain

Hyde Park

Chelsea Physic Garden

Green Spaces

Garden
120

Pastmans Park

Hoxton Square

Victoria Embankment Gardens

Southbank Centre

Southwark Park

N
W — E
S

Battersea Park

Barcelona Map (←)

Sevenoaks Map / Client: Sevenoaks So Much More / Design: Vicky Scott, Pillory Barn (www.moresevenoaks) (↑)

MÉDA
DE
CORO

SIERRA DE
PERIJA

SIERRA
SAN LUIS

LAGO
DE
MARACAIBO

GUARAMACAL

CIENAGA DEL
CATATU

RAYO DEL
CATATUMBO

SIERRA
NEVADA

PICO BOLIVAR

COLOMBIA

GABRIELA ACOSTA

Gabriela Acosta is a Strasbourg-based freelance
illustrator who studied architecture as a student.
She was originally from Maracaibo, a city where
it's summer almost all year long.

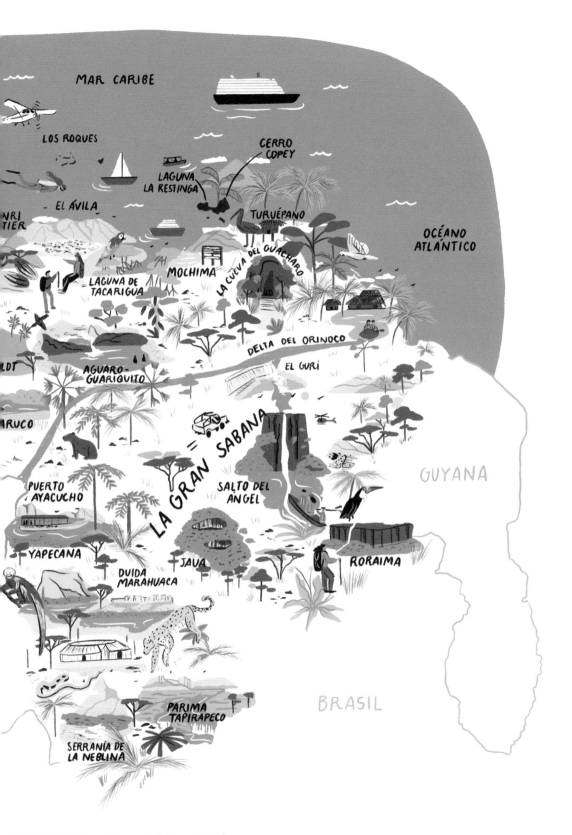

PAISAJES DE VENEZUELA
2022
GABRIELA ACOSTA

Mapa de Paisajes Naturales Venezuela ↖

PAISAJES DE ARGENTINA
2022
GABRIELA ACOSTA

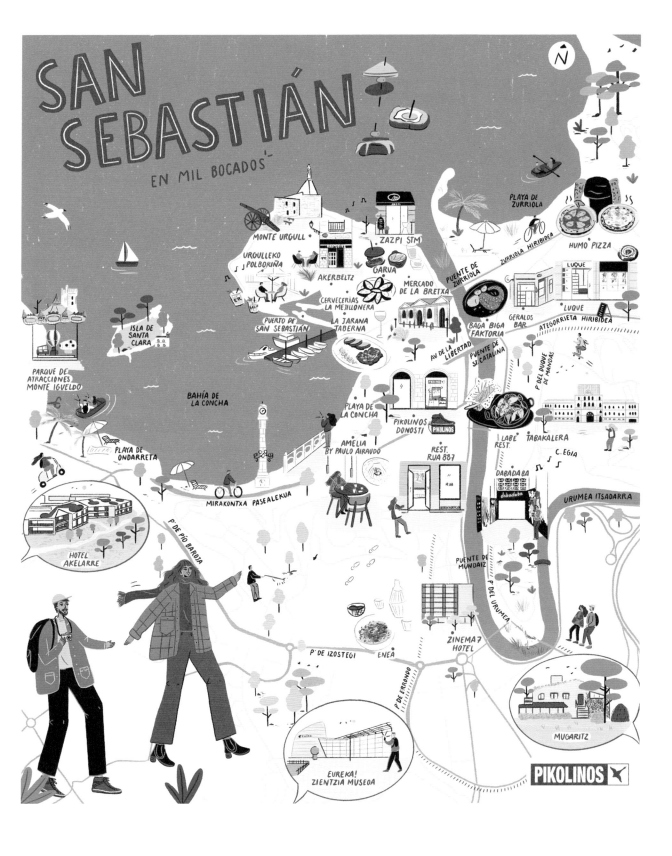

Smiling Cities de Pikolinos (La Barcelona) (←) (San Sebastián) (↖)
Client: Pikolinos

Cannes, France | Wonderful World of Maps (2021) (↗)
Client: Hotel Group Barrière

204

nes

HÔTEL BARRIÈRE LE GRAY D'ALBION

RUE DES SERBES

RUE D'ANTIBES

Hôtel Barrière
Le Gray d'Albion

Hôtel Barrière
Le Majestic

BOULEVARD DE LA CROISETTE

Majestic   Mademoiselle Gray
Plage Barrière

ÎLE SAINTE-MARGU

Lérins Isl

NIK NEVES

Berlin-based artist Nik Neves reflects his love for travel in his work. He crafts distinct visual stories across mediums such as illustrated maps, sketchbooks, and comics. Collaborating with Nina de Camargo on the Wonderful World of Maps project, their clients span Europe and North America.

Vila Madalena (Brazil) / Client: Dabba (2017)

New Orleans (USA) City Map (↖)
Client: Spartina (2018)

N
W E
S

150 meter
2 minutes walking
(1km is 4 minutes cycling)

NOORDERHAVEN

BUTJESSTRAAT

OUDE EBBINGESTRAAT

OUDE KIJK IN 'T JATSTRAAT

BROERPLEIN

VISSERS BRUG

ZWANESTRAAT

GULDENSTRAAT

HOGE DER A

STOELDRAAIERSTRAAT

VISMARKT

BRUGSTRAAT

HADDINGESTRAAT

FOLKINGESTRAAT

MUNNEKEHOLM

GE

EMMAPLEIN

UBBO EMMIUSSINGEL

STATIONSWEG

JAIME JACOB & TSJISSE TALSMA

Freelance illustrators Jaime Jacob and Tsjisse Talsma are part of Knetterijs, a group of 8 illustrators sharing a collective studio in Groningen. They work on commissions for clients at home and abroad across a wide variety of media.

PATRICIA BOLAÑOS

Patricia Bolaños, an illustrator based in New York, traded her architecture tools for coloured pencils after graduating from school in Madrid. Her illustrations grace the pages of Vogue, Harper's Bazaar, and Vanity Fair, while her collaborators include renowned brands like DKNY, Giorgio Armani, and Lancôme. In 2022, she published her debut book, New York is the Thing, and frequently sketches people, buildings, and street scenes.

New York City Guide / Client: Julie Flamingo (←)
La Vie est Belle / Client: Conde Nast Traveler (Spain), Lancôme (↖)

TING TSENG

Sunny Tseng is an illustrator and designer specialising in landscape architecture. His artwork is a harmonious blend of nature and culture, with the aim of resonating uniquely with viewers. Through his creations, he hopes to inspire a fresh perspective and appreciation for the environment.

TING TSENG

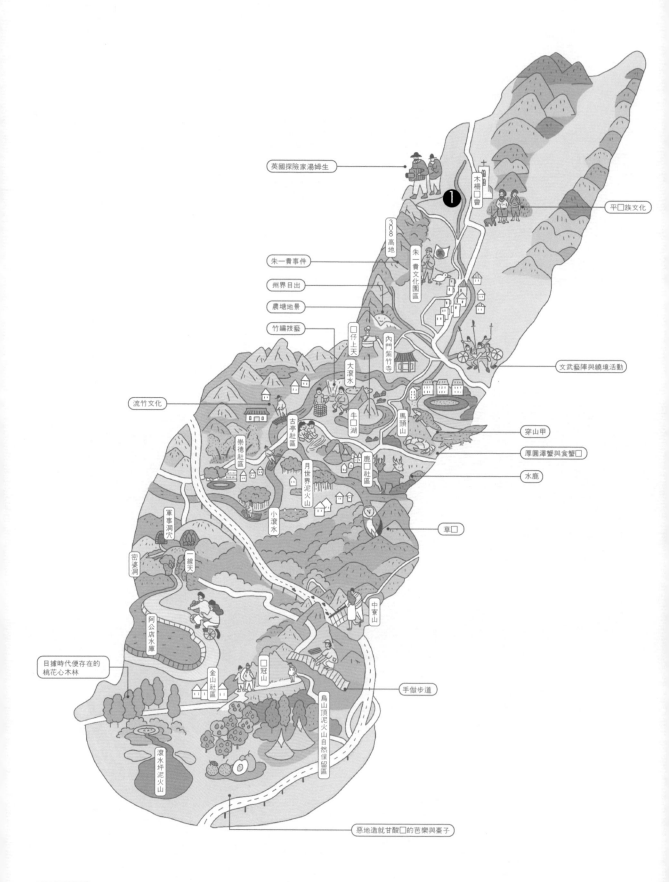

英國探險家湯姆生

土□□
木柵□會

平□族文化

朱一貴事件

308高地

朱一貴文化園區

州界日出

農塘地景

竹編技藝

□仔上天

內門紫竹寺

文武藝陣與繞境活動

大滾水

流竹文化

牛□湖

穿山甲

古亭社區

馬頭山

厚圓澤蟹與食蟹□

崇德社區

月世界泥火山

鹿□社區

水鹿

小滾水

軍事洞穴

草□

密婆洞

一線天

阿公店水庫

中寮山

日據時代便存在的
桃花心木林

金山社區

冠山

手做步道

烏山頂泥火山自然保留區

滾水坪泥火山

惡地造就甘酸□的芭樂與棗子

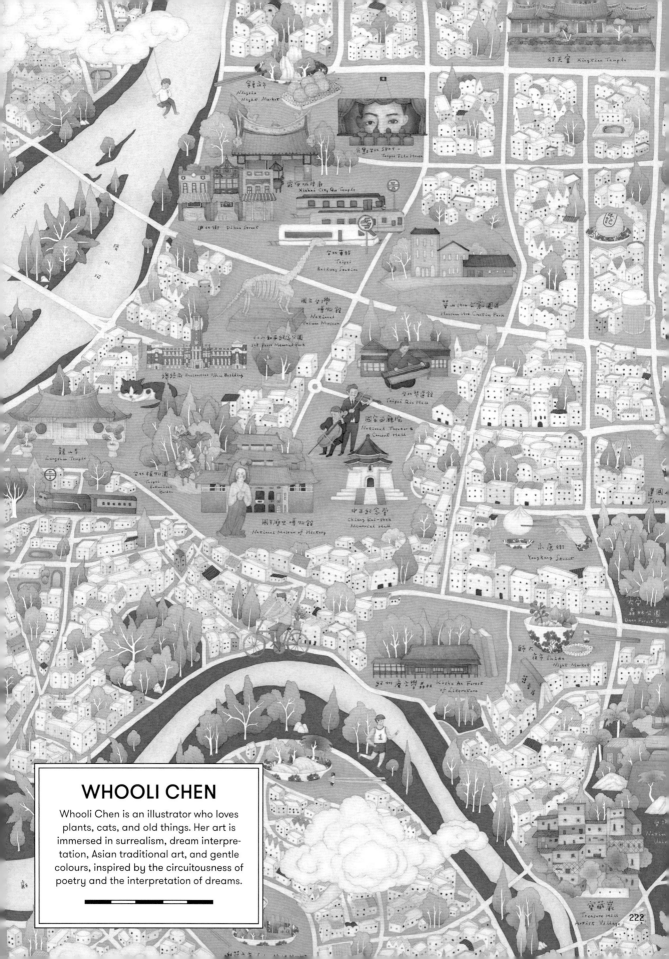

WHOOLI CHEN

Whooli Chen is an illustrator who loves plants, cats, and old things. Her art is immersed in surrealism, dream interpretation, Asian traditional art, and gentle colours, inspired by the circuitousness of poetry and the interpretation of dreams.

222

CITIX60: Taipei City Guide [大]

ESTADIO DEL CLUB INTERNACIONAL

LOSA DEPORTIVA

IGLESIA DE SAN LÁZARO

VIOLIN

LOS CRISTALES

QUINTA VIVANCO

TEMPLO DE LA TERCERA ORDEN FRANCISCANA

MUNDO ALPACA

AVENIDA JUAN DE LA TORRE

SANTA CATALINA

PARQUE ALPACA

PARQUE GRAU

CALLE PUENTE GRAU

MONASTERIO DE SANTA CATALINA

CASA DE LA MONEDA

Puente Grau

PARQUE HÉROES NAVALES

CALLE UGARTE

MUSEO CASA MOR

Puente Bajo Grau

AVENIDA LA MARINA

LA COMUNA TEATRO Y CINE

VILLALBA

PARQUE ECOLÓGICO QUINTA SALAS

Río Chili

CALLE MORAL

MONASTERIO Y MUSEO DE LA RECOLETA

ESCUELA SUPERIOR DE ARTE CARLOS BACA FLOR

LA RECOLETA

AVENIDA LA MARINA

MONO BLANCO AVENTURA

CALLE PUENTE BOLOGNESI

CRUZ VERDE

CENTRO ARTE

VID LAND

BEATERIO

CALLE PALACIO VIEJO

SKATEPARK LA MARINA

Puente Bolognesi

Puente Consuelo

CALLE CONSUELO

Centro Histórico de Arequipa

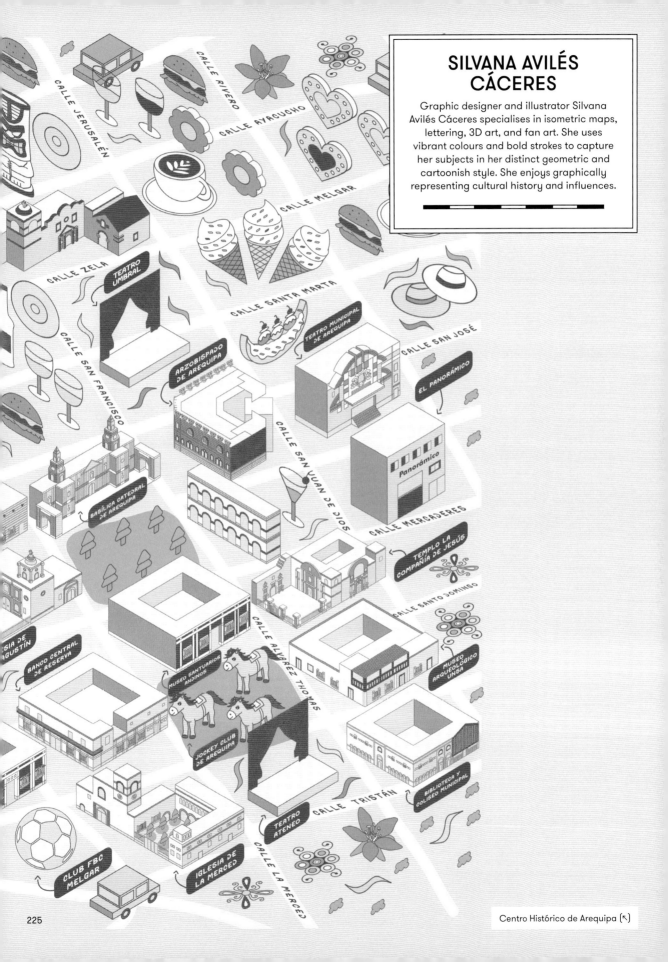

SILVANA AVILÉS CÁCERES

Graphic designer and illustrator Silvana Avilés Cáceres specialises in isometric maps, lettering, 3D art, and fan art. She uses vibrant colours and bold strokes to capture her subjects in her distinct geometric and cartoonish style. She enjoys graphically representing cultural history and influences.

MASAHIKO YONEMITSU

Masahiko Yonemitsu is an illustrator who specialises in creating imaginary towns and vision maps.

MASAHIKO YONEMITSU

230

OK's Village ↖

VIVIAN HO

Vivian has been active in the art and design scene since she graduated from Wesleyan University in economics and studio art, and she creates illustrations that open doors to worlds that are somehow familiar, yet maintain an exotic distance. She has held multiple solo and group exhibitions globally, and her work is collected by M+ Museum, the Valmont Foundation, Nishiji Collection and Copelouzos Family Art Museum.

Westergas Made With …,staat (↗)
Art Direction: Jordi Carles Subirà, Tom Schwaiger / Client: Westergas

PHILIP LINDEMAN

Across both commercial and personal projects, Philip Lindeman creates a fictional universe that reflects reality within his colourful and idiosyncratic illustrations. His work contains tragedies, mysticism, and comedic situations that feel recognisable from daily life and is filled with outlandish characters, details, and tantalising easter eggs.

PHILIP LINDEMAN

Werken Bij ProRail Made with Johnny Wonder (↑)
Art Direction: Jesse Reij / Client: ProRail

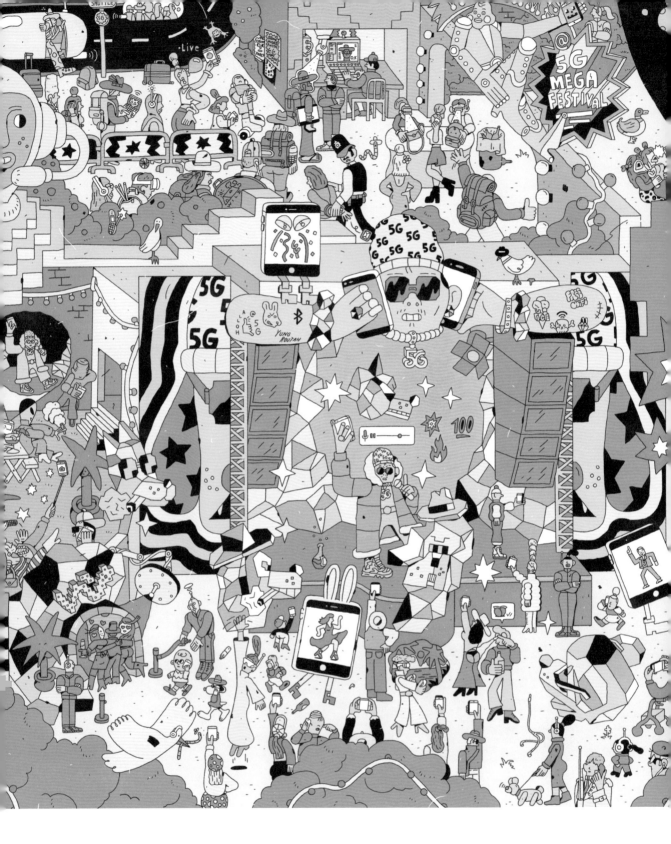

5G Mega Festival (↖)
Art Direction: Kara Melchers / Client: Canvas8

Stay Home, Stay Connected (↖)
Art Direction: Kara Melchers / Client: Canvas8

新界

丹桂坑

蓮花石澗

大嶼山

焦土海岸

香港島

鴨脷

大角

CARMEN NG

Carmen Ng is a contemporary artist whose paintings have been featured in numerous solo and group exhibitions. Through fine lines and a balanced composition, she combines traditional Chinese painting techniques with Western watercolours, while also experimenting with various textural effects.

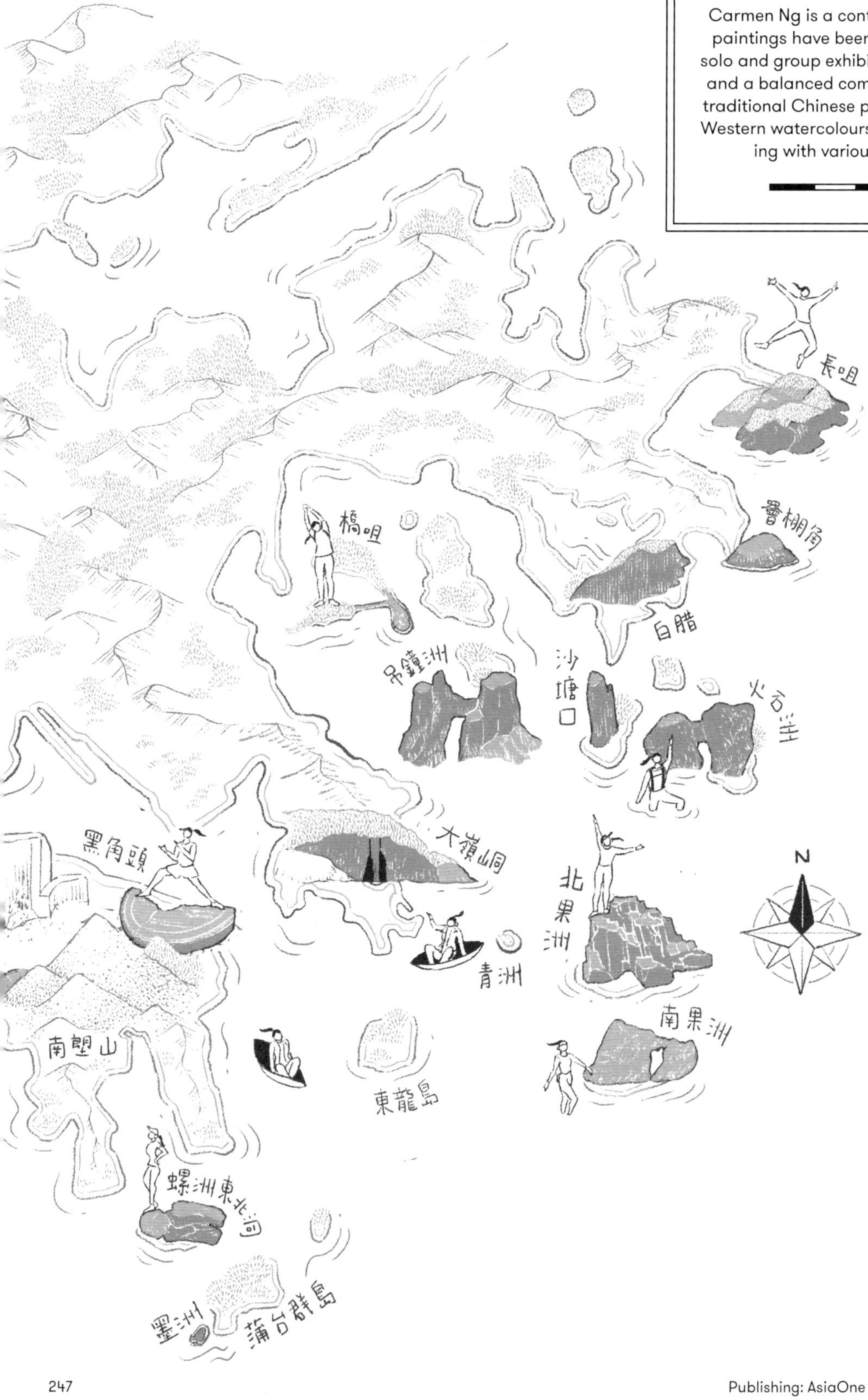
長咀
罾棚角
橋咀
白腊
吊鐘洲
沙塘口
火石洲
黑角頭
大嶺峒
N
北果洲
青洲
南果洲
南塱山
東龍島
螺洲東北洞
墨洲
蒲台群島

CAVE GIRL - Meko (↖)
Publishing: AsiaOne / Book Name: CAVE GIRL - Meko

LILA RUBY KING

Lila Ruby King is an independent illustrator who majored in printmaking. She is currently based in Athens, where she creates designs for the tourism industry. Today, her focus lies in working on maps of the Greek islands featuring the local architecture, food, and history.

Athens Map (Two Colour Edition) [人]

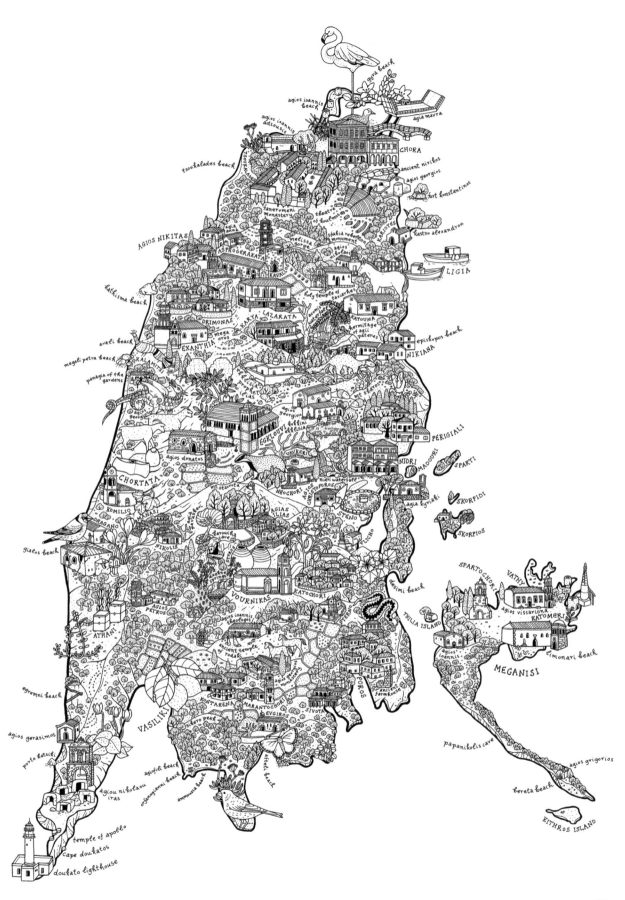

ACKNOWLEDGEMENTS

We would like to specially thank all the artists and
illustrators who are featured in this book for their significant
contribution towards its compilation. We would also like
to express our deepest gratitude to our producers for their
invaluable advice and assistance throughout this project,
as well as the many professionals in the creative industry
who were generous with their insights, feedback, and
time. To those whose input was not specifically credited or
mentioned here, we also truly appreciate your support.

FUTURE EDITIONS

If you wish to participate in viction:ary's future projects and
publications, please send your portfolio to:
we@victionary.com